Diophantus
and
Diophantine Equations

Cover design by Barbieri & Green

Library of Congress Catalog Card Number 97-74342
ISBN 0-88385-526-7

Printed in the United States of America

Current Printing (last digit):
10 9 8 7 6 5 4 3 2 1

Dolciani Mathematical Expositions

NUMBER TWENTY

Diophantus and Diophantine Equations

I. G. Bashmakova

Updated by Joseph Silverman
Translated from the Russian by Abe Shenitzer
with the editorial assistance of Hardy Grant

Russian original published by *Nauke,* Moscow, 1972

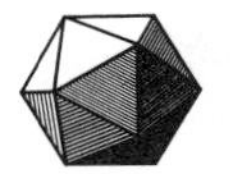

Published by
THE MATHEMATICAL ASSOCIATION OF AMERICA

THE
DOLCIANI MATHEMATICAL EXPOSITIONS

Published by
THE MATHEMATICAL ASSOCIATION OF AMERICA

The DOLCIANI MATHEMATICAL EXPOSITIONS series of the Mathematical Association of America was established through a generous gift to the Association from Mary P. Dolciani, Professor of Mathematics at Hunter College of the City University of New York. In making the gift, Professor Dolciani, herself an exceptionally talented and successful expositor of mathematics, had the purpose of furthering the ideal of excellence in mathematical exposition.

The Association, for its part, was delighted to accept the gracious gesture initiating the revolving fund for this series from one who has served the Association with distinction, both as a member of the Committee on Publications and as a member of the Board of Governors. It was with genuine pleasure that the Board chose to name the series in her honor.

The books in the series are selected for their lucid expository style and stimulating mathematical content. Typically, they contain an ample supply of exercises, many with accompanying solutions. They are intended to be sufficiently elementary for the undergraduate and even the mathematically inclined high-school student to understand and enjoy, but also to be interesting and sometimes challenging to the more advanced mathematician.

---

1. *Mathematical Gems,* Ross Honsberger
2. *Mathematical Gems II,* Ross Honsberger
3. *Mathematical Morsels,* Ross Honsberger
4. *Mathematical Plums,* Ross Honsberger (ed.)
5. *Great Moments in Mathematics* (*Before 1650*), Howard Eves
6. *Maxima and Minima without Calculus,* Ivan Niven
7. *Great Moments in Mathematics* (*After 1650*), Howard Eves
8. *Map Coloring, Polyhedra, and the Four-Color Problem,* David Barnette
9. *Mathematical Gems III,* Ross Honsberger
10. *More Mathematical Morsels,* Ross Honsberger
11. *Old and New Unsolved Problems in Plane Geometry and Number Theory,* Victor Klee and Stan Wagon
12. *Problems for Mathematicians, Young and Old,* Paul R. Halmos
13. *Excursions in Calculus: An Interplay of the Continuous and the Discrete,* Robert M. Young
14. *The Wohascum County Problem Book,* George T. Gilbert, Mark Krusemeyer, and Loren C. Larson

15. *Lion Hunting and Other Mathematical Pursuits: A Collection of Mathematics, Verse, and Stories by Ralph P. Boas, Jr.,* edited by Gerald L. Alexanderson and Dale H. Mugler
16. *Linear Algebra Problem Book,* Paul R. Halmos
17. *From Erdős to Kiev: Problems of Olympiad Caliber,* Ross Honsberger
18. *Which Way Did the Bicycle Go? ... and Other Intriguing Mathematical Mysteries,* Joseph D. E. Konhauser, Dan Velleman, and Stan Wagon
19. *In Pólya's Footsteps: Miscellaneous Problems and Essays,* Ross Honsberger
20. *Diophantus and Diophantine Equations,* I. G. Bashmakova

MAA Service Center
P. O. Box 91112
Washington, DC 20090-1112
800-331-1MAA FAX 301-206-9789

Dedicated to the memory of my husband

Andrei Ivanovich Lapin

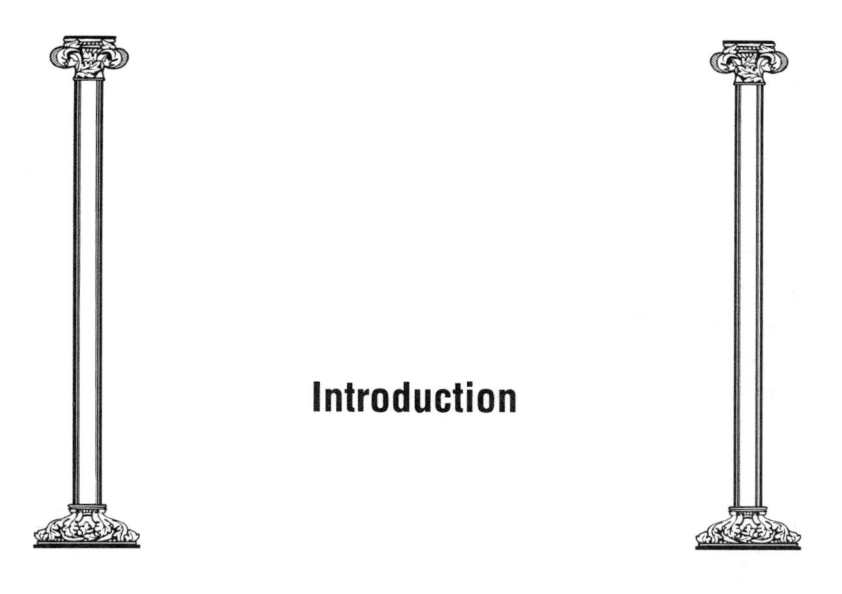

Introduction

Today, amateurs and professional mathematicians alike know about diophantine equations and even about diophantine analysis. In the second half of the 20th century this area of mathematics has become fashionable due to its proximity to algebraic geometry, an apparent focus of mathematical thought. Surprisingly, virtually nothing has been written about Diophantus, whose name is attached to indeterminate analysis and who is one of the most interesting scholars of antiquity. Even historians of mathematics have a fundamentally distorted view of his work. Most of them think that he solved particular problems, equivalent to indeterminate equations, by means of artful, particular methods. We will discuss such evaluations of the work of Diophantus in detail in Chapter 4.

Even a simple analysis of Diophantus' problems shows that he not only posed the problem of finding rational solutions of indeterminate equations but also gave some general methods for obtaining them. Here one must bear in mind that in the mathematics of antiquity general methods were never presented in "pure form," apart from the relevant problems. For example, when Archimedes determined the area of an ellipse, of a parabolic segment, of a sphere, the volume of a sphere and of other solids, he used the method of integral sums and of passage to the limit without ever giving a general, abstract description of these methods. 17th- and 18th-century scholars had to study carefully and reinterpret his works in order to distill from them his methods. The same applies to Diophantus. His methods were understood and applied to the solution of new problems by Viète and Fermat at the time when Archimedes' works were being deciphered. In our investigations we will follow the example of Viète and Fermat: we will analyze the solutions of concrete problems in order to understand the general methods involved.

We wish to add that whereas the discovery of the integral and differential calculus by Newton and Leibniz basically brought to a close the evolution of Archimedes' integral methods, the evolution of Diophantus' methods extended for a few more centuries and interlaced with that of the theory of algebraic functions and of algebraic geometry. The evolution of the ideas of Diophantus can be traced all the way to the works of Henri Poincaré and André Weil. This makes the history of diophantine analysis especially interesting.

This book is largely concerned with Diophantus' methods of obtaining rational solutions of indeterminate equations of the second and third orders and with their history. In passing, we will consider the question of the number system used by Diophantus and his literal symbolism. Even this far simpler question has not yet been cleared up. Most historians of science are of the opinion that Diophantus limited himself to positive rational numbers and knew no negative numbers. We will try to show that this is not the case, that in his "Arithmetic" Diophantus extended the domain of numbers to the field Q of rational numbers.

I hope that this book will introduce the reader to a new aspect of the mathematics of antiquity. The view most of us have of this mathematics is based on Euclid's "Elements" and on the works of Archimedes and Apollonius. Diophantus opens before us the equally rich and beautiful world of arithmetic and algebra.

Of course, we cannot deal with all of Diophantus' works, and certainly not with all of diophantine analysis and its history. As we said earlier, we will restrict ourselves basically to the area known as the arithmetic of algebraic curves. This area deals with the finding of the rational points on such curves (or equivalently, with finding the rational solutions of single algebraic equations in two unknowns) and with the study of their structure. That is why the reader will not find here the history of the problem of finding integer solutions of indeterminate equations, a problem studied by Fermat, Euler, Lagrange, and Legendre, as well as by modern mathematicians. Nor will we study the difficult and subtle problem of the existence of rational (or integral) solutions of indeterminate equations with rational integral coefficients, for this problem lies outside the circle of problems directly related to Diophantus. One more disclaimer: we will not deal with the history of Hilbert's tenth problem, devoted to finding a general method (or proving that none exists) for deciding in a finite number of steps whether or not a given equation with rational integral coefficients has a solution.

This book is intended for a broad public. It can be read by high school and university teachers, by students in physical-mathematical faculties of universities and pedagogical institutes, by engineers, and by seniors in mathematically oriented high schools. Strictly speaking, the reading of the book requires only the knowledge of analytic geometry and of the elements of the differential and integral calculus. This means that not all of the chapters in this book are equally accessible to high school students. To make it easier to use the book we provide a number of explanatory remarks which describe the structure of the book and indicate which chapters can be omitted without affecting the understanding of the book as a whole.

Chapter 1 is devoted to Diophantus as a person and Chapter 2 to the system of numbers and symbols he introduced. Chapter 3 contains information about diophantine equations and algebraic geometry indispensable for understanding the rest of the book. Chapter 4 deals with evaluations of the methods of Diophantus by historians of mathematics. In Chapters 5 and 6 we present Diophantus' problems and study the methods he used to solve indeterminate equations of second and third order. We also explain to the reader the matter of homogeneous, or projective, coordinates. In Chapter 7 we present some of Diophantus' problems which required number-theoretic investigations. These problems enable one to judge how much the ancient mathematicians knew about number theory. The rest of the book, that is Chapters 8–13, is devoted to the history of Diophantus' methods from the time of Viète and Fermat until the 1920s. In Chapter 10 we talk about the theorem which involves the addition of Euler's elliptic integrals and about its use by Jacobi for finding the rational points on a cubic curve. The reader of this chapter must be familiar with the concept of an improper integral. High school students can omit this material. But then they should also omit the first two paragraphs of Chapter 11. In Chapters 12 and 13, devoted to the relevant works of Henri Poincaré and to certain subsequent results, many questions have been presented in a sketchy manner and some, requiring the introduction of new and complex concepts, have been omitted altogether. Nevertheless, I am confident that the reader will get an idea of the works of Diophantus and of the history of the arithmetic of algebraic curves and may possibly take an interest in this most beautiful area of mathematics.

I wish to offer my deep thanks to A.I. Lapin and I.P. Shafarevich for many valuable remarks and hints, to the editor N.N. Hendrikson for improvements and corrections in the manuscript, to A. Shenitzer for his wonderful translation of the book that gives it new life, and to J.H. Silverman for his notes on the history of algebraic geometry in the last twenty years.

At the end of the book I give a list of the most accessible editions of Diophantus "Arithmetic" and of works about it.

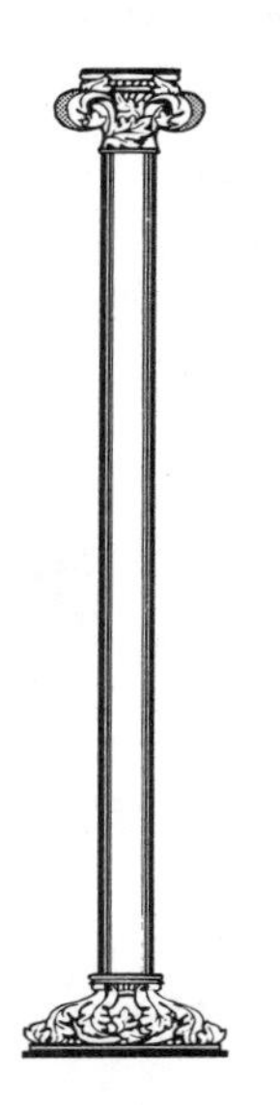
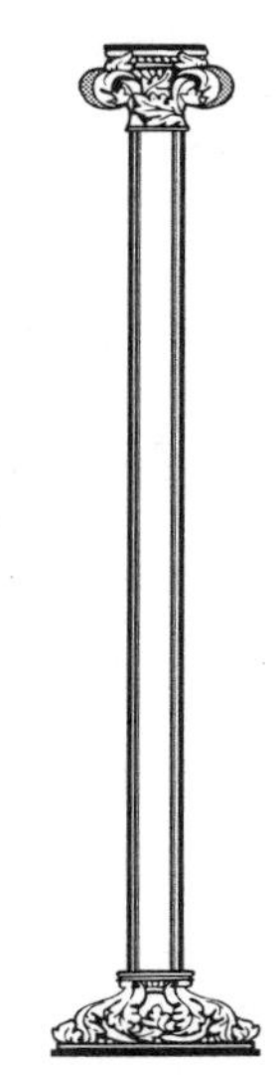

Contents

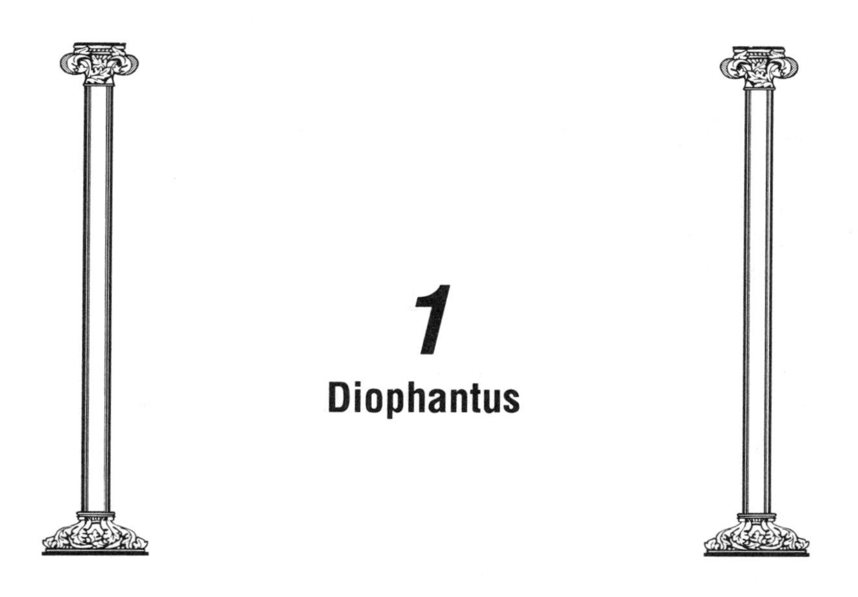

1 Diophantus

Diophantus represents one of the most difficult riddles in the history of science. We do not know when he lived and we do not know his predecessors who may have worked in the same area. His works resemble a fire flashing in an impenetrable darkness.

He may have lived at any time during a 500-year period! Its lower bound is easy to determine. In his book on polygons Diophantus frequently mentions the mathematician Hypsicles of Alexandria who lived in the second century BC. On the other hand, Theon of Alexandria's commentary on the "Almagest" of the great astronomer Ptolemy contains an excerpt from Diophantus' work. Theon lived in the middle of the fourth century AD. Hence the 500-year period!

The French historian of science Paul Tannery, editor of the most complete text of Diophantus, tried to narrow this time interval. In the Escurial library he found excerpts from the writings of Michael Psellus, a Byzantine scholar of the 11th century, which state that "the most learned Anatolii, having collected the most essential parts of this science (a reference to the introduction of powers of the unknown and of the corresponding notation), dedicated them to his friend Diophantus." Anatolii of Alexandria actually wrote an "Introduction to Arithmetic," and excerpts from this work are quoted in preserved works of Iamblichus and Eusebius. Now Anatolii lived in Alexandria in the middle of the third century AD, more precisely until 270, when he became the bishop of Laodicea. This means that his friendship with Diophantus, invariably referred to as Diophantus of Alexandria, must have preceded this date. Thus if the famous Alexandrian mathematician and Anatolii's friend named Diophantus are the same person, then Diophantus must have lived in the middle of the third century AD.

Diophantus' "Arithmetic" is dedicated to the "reverend Dionysus" who, as follows from the "Introduction," was interested in arithmetic and its teaching. While the name Dionysus was relatively common at the time, Tannery assumed that the "reverend" Dionysus had to be sought among the well-known people of that period who occupied prominent positions. Surprisingly enough, it turned out that a certain Dionysus, who from 231 on was the director of Alexandria's Christian high school, became the city's bishop in 247! That is why Tannery identified this Dionysus with the one Diophantus dedicated his work to, and thus arrived at the conclusion that Diophantus lived in the middle of the third century. In the absence of a better chronology, Tannery's will do[1].

But the place where Diophantus lived is well known. It is the famous Alexandria, center of the scientific thought of the Hellenic world.

Following the breakup of Alexander the Great's huge empire, Egypt was ruled by Ptolemy Lagus, one of Alexander's generals, who made the new city of Alexandria its capital. This multilingual commercial center soon became one of the most beautiful cities of antiquity. Eventually Rome surpassed it in size, but for a long time it had no equal. For many centuries, this city was the scientific and cultural center of the ancient world, because of Ptolemy's founding of the Musaeon, the temple of the Muses, a kind of Academy of Sciences which attracted leading scholars. These scholars were paid salaries and their duty was essentially to meditate and to engage in discussions with their students. It included a splendid library which at one time had as many as 700,000 manuscripts. Small wonder that scholars and young men thirsting for knowledge flocked to Alexandria to listen to distinguished philosophers, to learn astronomy and mathematics, and to immerse themselves in the study of unique manuscripts in the cool rooms of the library.

From the third to the second century BC the Musaeon shone with the names of Euclid, Apollonius, Eratosthenes, and Hipparchus. In the early centuries BC it suffered a temporary decline, due to the decline of the house of the Ptolemies and to the Roman conquests (Alexandria was conquered in 31 AD), but in the early centuries AD it was regenerated owing to the support of the Roman emperors. From the first to the third century scholars such as Heron, Ptolemy, and Diophantus worked here. Alexandria continued to be the world's scientific center. In this respect Rome was never its rival. There simply was no such thing as Roman science, and the Romans remained true to the legacy of Vergil, who wrote:

> Let others fashion from bronze more lifelike, breathing images—
> For so they shall—and evoke living faces from marble;
> Others excel as orators, others track with their instruments
> The planets circling in heaven and predict when stars will appear.
> But, Romans, never forget that government is your medium!

In order to make use of all that is known about the person of Diophantus, we quote the following riddle:

> God granted him to be a boy for the sixth part of his life, and adding a twelfth part to this, He clothed his cheeks with down; He lit him the light of wedlock after a seventh part, and five years after his marriage He granted him a son. Alas! late-born wretched child; after attaining the measure of half his father's life, chill Fate took him. After consoling his grief by this science of numbers for four years he ended his life.

From this it is easy to compute that Diophantus lived to be 84 years old. But for this there is no need to master the art of Diophantus. It suffices to be able to solve a first-order equation in one unknown, something the Egyptian scribes could do in 2000 BC.

But the most mystifying riddle is the works of Diophantus. Only six of the 13 books which make up the "Arithmetic"[2] have come down to us. Their style and contents differ radically from the classical ancient works on number theory and algebra whose models we know from Euclid's

"Elements" and his "Data" and from the lemmas of Archimedes and Apollonius. The "Arithmetic" is undoubtedly the result of numerous investigations which are completely unknown to us. We can only guess at its roots and admire the richness and beauty of its results.

Diophantus' "Arithmetic" is a collection of problems (189 in all) each of which comes with one or more solutions and the necessary explanations. Hence the initial impression that this is not a theoretical work. But a thorough perusal shows that the problems have been carefully selected, and serve to illustrate definite, rigorously thought-out methods. Following the norm of antiquity, the methods are not stated in general form but reappear in the solutions of problems of the same type.

But the first book is preceded by the author's "general introduction," on which we will dwell at some length.

Notes

[1] Raphael Bombelli, who familiarized himself with the manuscript of the "Arithmetic" in the Vatican library around 1570, wrote with assurance that Diophantus had lived at the time of Antoninus Pius, that is, in the middle of the second century AD. Fermat also held to this date.

[2] In the introduction, Diophantus states that the "Arithmetic" is divided into 13 books. Only six books in old Greek have come down to us. In 1972 there came to light an Arabic manuscript allegedly containing books four, five, six, and seven. These books do not coincide with the corresponding Greek versions. In 1974 R. Rashed published a French translation of the manuscript and in 1975 the text itself. In the same year J. Sesiano published a critical version of the text as well as an English translation with detailed commentaries. It seems that these books were part of another, probably later, version of the "Arithmetic" that has not come down to us. In particular, the manuscript contains the 8th and 9th powers of the unknown, whereas the only powers introduced by Diophantus are nth powers with $-6 \leq n \leq 6$.

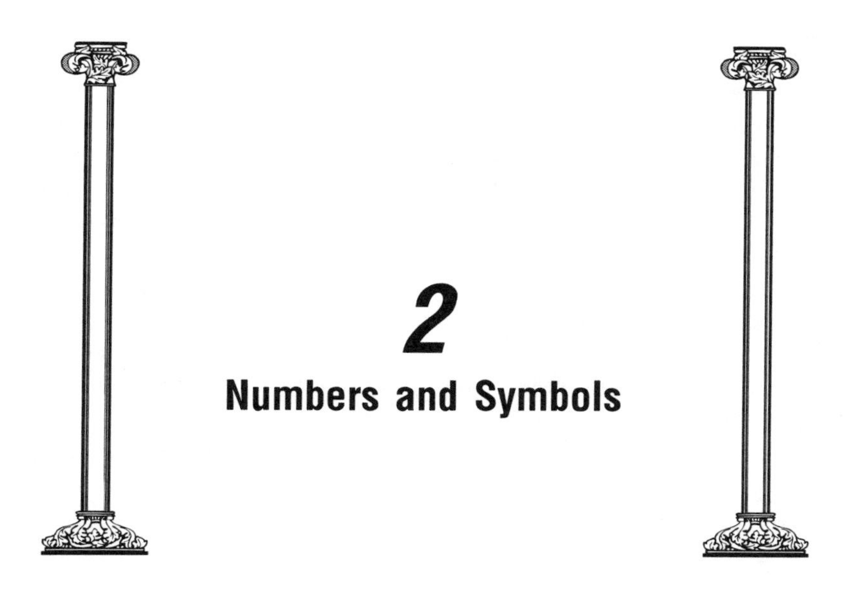

2 Numbers and Symbols

Diophantus begins with fundamental definitions and a description of the literal symbols which he will use.

In classical Greek mathematics, which reached its completion in Euclid's "Elements," number (*ἀριθμός*; hence the name "arithmetic" for the science of numbers) meant a collection of units, that is a whole number. Fractions and irrational quantities were not called numbers. Strictly speaking, there are no fractions in the "Elements." Unity was viewed as indivisible, and instead of fractions of unity one considered ratios of whole numbers. Irrational quantities took the form of ratios of incommensurable segments. Thus for the Greeks of the classical period the number which is now denoted by $\sqrt{2}$ was the ratio of the diagonal of a square to its side. There were no negative numbers. Nor were there equivalents of negative numbers. In the case of Diophantus the picture is radically different.

Diophantus gives the traditional definition of a number as a collection of units, but when it comes to his problems he looks for *positive rational* solutions and calls each of them a *number* (*ἀριθμός*).

But that is not all. Diophantus introduces negative numbers: he refers to them by the special term *λεῖψις*, derived from the word *λεῖπω* - to be missing, not to suffice—so that the term itself could be rendered as "shortage." Diophantus calls a positive number ὕπαρξις—existence, being. The plural form of this word denotes property. Thus Diophantus' terms for signed numbers are close to those used during the Middle Ages in the East and in Europe. Most likely, these terms were simply translations from the Greek into Arabic, Sanskrit, and Latin, and then into the different European languages.

Many translators of Diophantus render λεῖψις as subtrahend. This is wrong. In fact, to indicate the operation of subtraction Diophantus uses the terms ἀφελεῖν or ἀφαιρεῖν, derived from the word ἀφαιρέω, to subtract. When transforming equations Diophantus frequently uses the standard expression "we will add to both sides λεῖψις."

We have gone into a detailed philological analysis of Diophantus' text in order to convince the reader that it is correct to translate Diophantus' terms as "positive" and "negative."

Diophantus formulates for relative numbers the following rule of signs:

> a negative multiplied by a negative yields a positive, whereas a negative by a positive yields a negative, and the distinctive sign for the negative is ⋔, an inverted and shortened (letter) ψ.

He goes on:

> Now that I have explained to you multiplication of powers and their reciprocals, division of such expressions likewise becomes clear. It will now be a good thing for the beginner to do exercises involving addition, subtraction, and multiplication of algebraic expressions. He must know how to add positive as well as negative expressions with different coefficients to other expressions, which may be positive or, equally, positive as well as negative, and to subtract from expressions which may be sums or differences other magnitudes which may themselves be sums or differences.

Note that while Diophantus is looking only for positive rational solutions, he readily uses negative numbers in intermediate computations. Thus it is safe to say that Diophantus extended the domain of numbers to the field of rationals, where one can easily carry out all four arithmetical operations.

In the "Arithmetic" we encounter for the first time literal symbolism. Diophantus introduced the following notations for the six powers $x, x^2, \ldots, x^6$ of the unknown x:

the first power—ς;
the second power—$\Delta^{\tilde{\upsilon}}$, from $\Delta\acute{\upsilon}\nu\alpha\mu\iota\varsigma$, force, power;
the third power—$\mathrm{K}^{\tilde{\upsilon}}$, from $\mathrm{K}\acute{\upsilon}\beta o\varsigma$, cube;
the fourth power—$\Delta^{\tilde{\upsilon}}\Delta$, from $\Delta\acute{\upsilon}\nu\alpha\mu o\delta\acute{\upsilon}\nu\alpha\mu\iota\varsigma$, square square;
the fifth power—$\Delta\mathrm{K}^{\tilde{\upsilon}}$, from $\Delta\acute{\upsilon}\nu\alpha\mu o\chi\acute{\upsilon}\beta o\varsigma$, square cube;
the sixth power—$\mathrm{K}^{\tilde{\upsilon}}\mathrm{K}$, from $\mathrm{K}\acute{\upsilon}\beta o\chi\acute{\upsilon}\beta o\varsigma$, cube cube.

Diophantus denotes the constant term, that is x^o, by the symbol $\overset{\circ}{\mathrm{M}}$, that is by the first two letters in $\mu o\nu\acute{\alpha}\varsigma$, or unity.

He introduced a special symbol $^{\chi}$ for negative exponents. In this way he could denote the first six negative exponents of the unknown. For example, he denoted x^{-2} and x^{-3} by $\Delta^{\tilde{\upsilon}\chi}$ and $\mathrm{K}^{\tilde{\upsilon}\chi}$ respectively.

Thus Diophantus had a symbolism for denoting the positive and negative powers of a single unknown up to and including the sixth power. He failed to introduce a symbol for a second unknown, which greatly complicated the solution of problems. Sometimes, within a single problem, ς denoted more than one unknown number. In addition to these symbols, Diophantus used the symbol $\square$ for an indeterminate square. For example, if, by the condition of the problem, the sum of the product of two numbers and one of them is to be a square then the latter is denoted by $\square$.

Then Diophantus gives the rules for multiplication of x^m by x^n for positive and negative m and n ($|m| \leqq 6,\ |n| \leqq 6$).

For an equality sign Diophantus used the symbol $\mathring{\iota}'\sigma$—the first two letters in the word $\mathring{\iota}'\sigma o\varsigma$, which means equal. All this enables him to write equations in literal form. For example, he wrote the equation

$$202x^2 + 13 - 10x = 13,$$

more precisely

$$x^2 202 + x^0 13 - x 10 = x^0 13,$$

as

$$\Delta^{\tilde{\upsilon}}\overline{\sigma\beta}{}^{\circ}\overline{\iota\gamma} \pitchfork \varsigma\bar{\iota}\,\mathring{\iota}'\sigma\overset{\circ}{\mathrm{M}}\overline{\iota\gamma}.$$

The Greeks used the letters of the alphabet with bars over them to denote numbers. The first nine letters $\overline{\alpha}, \overline{\beta}, \dots, \overline{\vartheta}$ denoted the numbers from 1 to 9. The next nine denoted the multiples of 10 from 10 to 90. The last nine (the alphabet of 24 letters was augmented by the addition of three older letters) denoted the nine hundreds. Thus, for example, $\overline{\sigma} = 200$, $\overline{\beta} = 2$, so that $\overline{\sigma\beta}$ denoted 202. Similarly, $\overline{\iota} = 10$, $\overline{\gamma} = 3$, so that $\overline{\iota\gamma} = 13$.

In the "Introduction" Diophantus formulated rules of transformation of equations which involved addition of equal terms to both sides of the equation and reduction of like terms. Later, these two rules became well known under their Arabized names of *al-jabr* and *al-muqabala*.

We see that when it comes to naming and denoting powers of the unknown, Diophantus—like ourselves—uses the geometric terms "square" and "cube." But when setting down equations he calmly adds a square or a cube to a side [1], that is, he treats them not as geometric images but as numbers. Also, he finds it possible to introduce "square squares," "square cubes," and so on, without any thought of tying them to multidimensional spaces. In other words, in using geometric terminology Diophantus was merely following an established tradition.

Thus we encounter here a completely new construction of algebra, based on arithmetic and not, as in Euclid's case, on geometry. But far from being a simple return to the numerical algebra of the Babylonians, this is the beginning of a construction of literal algebra, which found its proper language in the works of Diophantus.

Notes

[1] In the so-called Greek geometric algebra, addition was defined only for homogeneous magnitudes, that is, one could add lengths to lengths and areas to areas but not a length to an area (not, say, the side of a square to the square). Addition was regarded as a geometric operation ("application") and not as the arithmetical operation of addition of the relevant numbers.

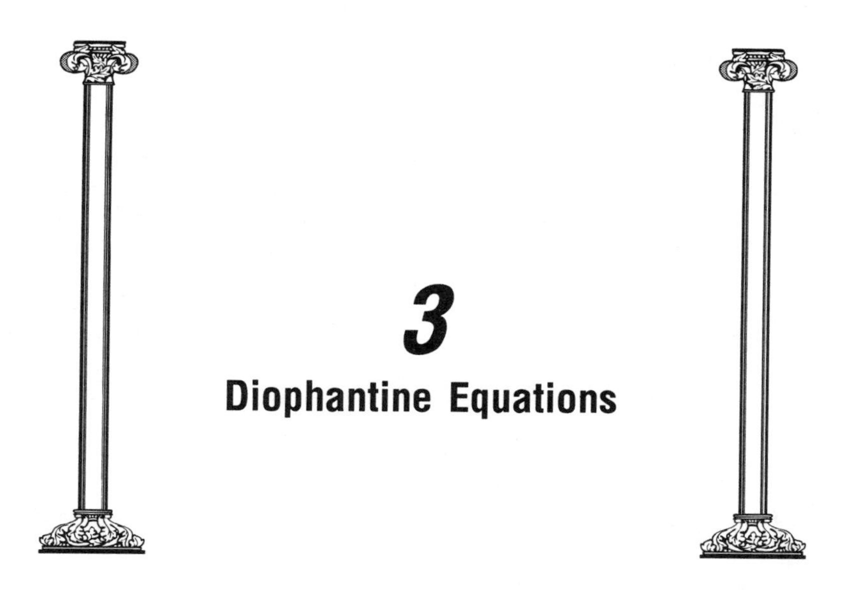

3 Diophantine Equations

What is most surprising about the "Arithmetic" is not only Diophantus' use of a completely new language and his bold extension of the domain of numbers but the problems he posed and solved.

To understand the essence of these problems and to investigate Diophantus' methods we must begin by providing some information from algebraic geometry and from the theory of indeterminate equations. At the present time, the problem of the solution of indeterminate equations is formulated as follows: given m polynomials in n variables, $m < n$, $f_1(x_1, x_2, \ldots, x_n), \ldots, f_m(x_1, x_2, \ldots, x_n)$ with coefficients in some field k,[1] find the set $M(k)$ of all rational solutions of the system

$$\begin{aligned} f_1(x_1, x_2, \ldots, x_n) &= 0, \\ &\cdots \\ f_m(x_1, x_2, \ldots, x_n) &= 0, \end{aligned} \tag{1}$$

and determine its algebraic structure. A solution $(x_1^{(o)}, \ldots, x_n^{(o)})$ is said to be *rational* if all the $x_i^{(o)} \in k$.

Of course, the set $M(k)$ depends on the field k. Thus, for example, the equation $x^2 + y^2 = 3$ has no solutions in the field Q of rational numbers but has infinitely many solutions in the field $Q(\sqrt{3})$, that is, in the set of numbers of the form $a + b\sqrt{3}$, where a and b are rational numbers [2].

The most important cases for number theory are the case $k = Q$ and the case when k is the field of residue classes mod p, p a prime. Diophantus considered the first of these cases and so will we.

We will limit ourselves to the consideration of Diophantus' problems which can be reduced to a single equation in two unknowns, that is to the case $m = 1, n = 2$:

$$f(x, y) = 0. \tag{2}$$

This equation determines in the plane $R^{(2)}$ an *algebraic curve* Γ. We will call a rational solution of (2) a *rational point* on Γ.

Nowadays, geometric language is an integral part of mathematical thinking. It makes it easier to explain and to understand many facts. This being so, we will frequently employ it in the sequel, although Diophantus never made use of it.

First it is necessary to give some classification of the equations (2) or, equivalently, of algebraic curves. The most natural, and historically earliest, such classification was the one by order.

The *order* of a curve (2) is the maximal order of the terms of the polynomial $f(x, y)$ (the order of a term is the sum of the exponents of x and y in the term). The geometric sense of this notion is that the number of points of intersection of a curve of order n and a straight line is n. When counting the number of these points one must consider their multiplicities and include complex points and "points at infinity" (see p. 27). Thus, for example, the circle $x^2 + y^2 = 1$ and the straight line $x + y = 2$ intersect in two complex points, the hyperbola $x^2 - y^2 = 1$ and the straight line $y = x$ in two points at infinity, and this same hyperbola and the straight line $x = 1$ in a single point of multiplicity 2.

From the viewpoint of *diophantine analysis* (this is the name of the branch of mathematics that grew out of the problem of solving indeterminate equations; now the more common name is *diophantine geometry*) classification by order has turned out to be rather crude. An example will clarify this assertion.

Consider the circle $C: \ x^2 + y^2 = 1$ and an arbitrary straight line with rational coefficients, say, $L: \ y = 0$. We will show that there is a one-to-one correspondence between the rational points on these two curves. One way of showing this is the following: fix the point $A(0, -1)$ on C and associate with each rational point B on L the point B' on C in which the straight line AB intersects C (Figure 1).

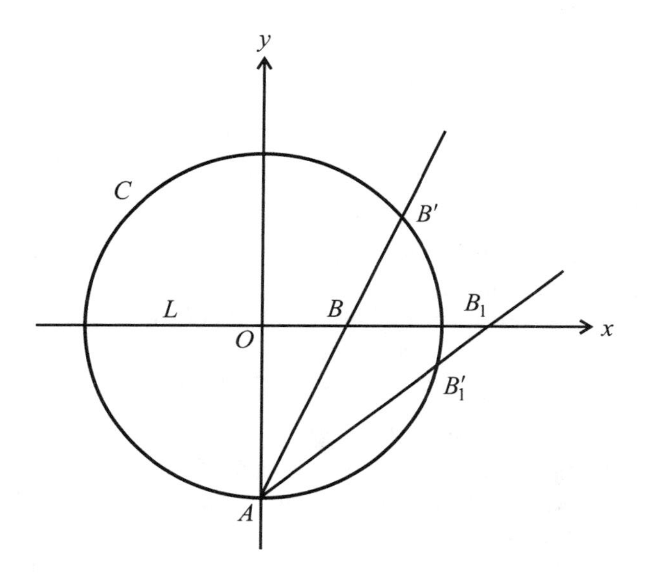

FIGURE 1

We leave it to the reader to prove that the coordinates of B' are rational or to read Diophantus' proof (presented in Chapter 5) of this fact. Obviously, it is possible to set up this kind of correspondence between the rational points on any conic section with a rational point and any rational straight line. This shows that from the viewpoint of diophantine analysis the circle C and the straight line L are indistinguishable. Their respective sets of rational points are equivalent in spite of the fact that their orders are different.

A finer classification of algebraic curves is by genus. This classification was introduced in the 19th century by Abel and Riemann and takes into consideration the singular points on a curve Γ.

We will assume that the polynomial $f(x, y)$ in equation (2) of Γ is *irreducible* over the rationals, that is, it cannot be written as a product of polynomials with rational coefficients. The tangent to Γ at a point $P(x_o, y_o)$ is given by

$$y - y_o = k(x - x_o),$$

where

$$k = -\frac{f_x(x_o, y_o)}{f_y(x_o, y_o)}.$$

If f_x or f_y is different from zero at P, then the coefficient k of the tangent has a definite value (if $f_y(x_o, y_o) = 0$ and $f_x(x_o, y_o) \neq 0$, then $k = \infty$ and the tangent at P is vertical).

If both partial derivatives vanish at P, that is, if

$$f_x(x_o, y_o) = 0 \quad \text{and} \quad f_y(x_o, y_o) = 0,$$

then the point P is said to be *singular*.

For example, the point $(0, 0)$ is a singular point of the curve $y^2 = x^2 + x^3$ because at this point both partial derivatives $f_x = -2x - 3x^2$ and $f_y = 2y$ vanish.

The simplest singular points are *double* points, at which at least one of the partial derivatives f_{xx}, f_{xy}, and f_{yy} does not vanish. In Figure 2 we see a double point at which the curve has two different tangents. More complex singular points are shown in Figure 3.

An algebraic curve can have at most a finite number of singular points. Indeed, let

$$f(x, y) = 0 \tag{$*$}$$

be the equation of the curve, where $f(x, y)$ is irreducible over Q. The coordinates of the singular points must satisfy the equations

$$f_x(x, y) = 0, \qquad f_y(x, y) = 0,$$

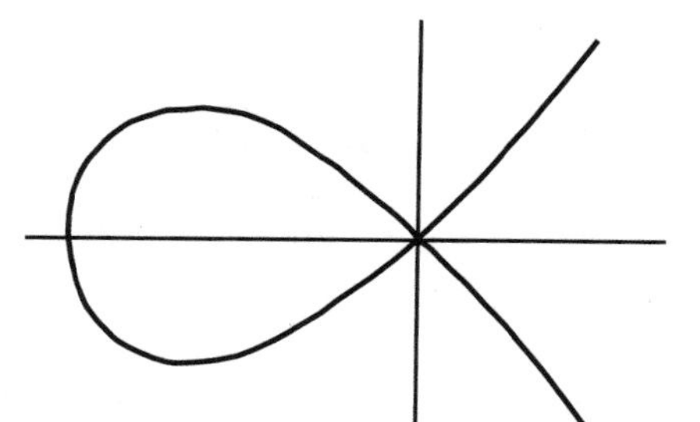

FIGURE 2

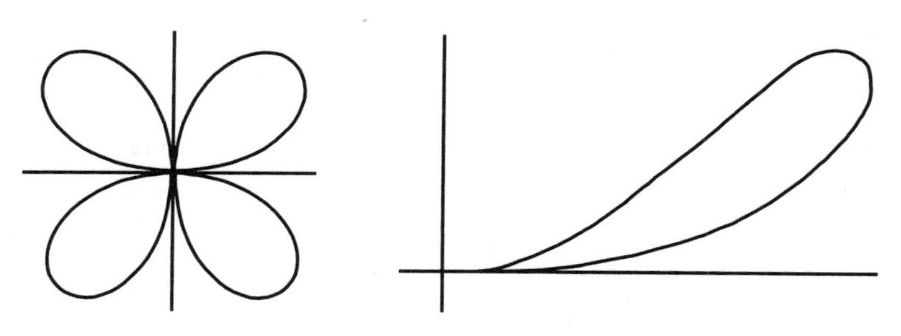

FIGURE 3

and the equation $(*)$. But the system of these three algebraic equations can have only a finite number of solutions.

We will now define the genus of a plane algebraic curve whose only singular points are double points. (In the general case, that is in the case of an arbitrary algebraic curve with arbitrary singularities, the definition is relatively complicated. We will not give this definition here because we will not need it.)

Thus let Γ be a plane algebraic curve with d double points ($d \geqq 0$). Then by the *genus* of Γ we mean the integer p defined by the formula

$$p = \frac{(n-1)(n-2)}{2} - d,$$

where n is the order of Γ. It is possible to show that $p \geqq 0$.

If Γ is a straight line or a second-order curve, then our formula yields for p the value 0, that is the curves are of the same genus. A third-order curve has genus 1 if it has no double points and genus 0 if it has one double point. For example, the so-called Fermat curve $x^3 + y^3 = 1$ has genus 1.

Classification by genus also fails to take into consideration the arithmetic properties of a curve. For example, the curves $x^2 + y^2 = 1$ and $x^2 + y^2 = 3$ have genus 0, but the first has infinitely many rational points and the second none. To find a classification of curves which is adequate from the viewpoint of diophantine analysis we note that when solving equation (2) we often make a change of variables

$$x = \varphi(u, v), \ \psi(u, v), \tag{3}$$

where φ and ψ are rational functions, that is quotients of polynomials. Substituting (3) in (2) we obtain

$$G(u, v) = 0. \tag{4}$$

This equation determines a certain curve Γ'. For the rational points on Γ, with the possible exception of finitely many of them, to go over into rational points on Γ' and, conversely, for the preimages of rational points on Γ' to be rational points on Γ, it is necessary and sufficient that the functions φ and ψ have rational coefficients and that the equations (3) be invertible, that is, that it should be possible to find rational functions φ_1 and ψ_1 with rational coefficients such that

$$u = \varphi_1(x, y), \quad v = \psi_1(x, y). \tag{3'}$$

If it is possible to establish a correspondence between curves Γ and Γ' by means of formulas of the type (3) and (3′) with rational coefficients, then the curves are said to be *birationally equivalent,* and the transformations are called *birational*.

For example, if $\varphi(u, v)$ and $\psi(u, v)$ are linear functions, that is if

$$x = \varphi(u, v) = au + bv + c,$$

$$y = \psi(u, v) = au_1 + b_1v + c_1,$$

and

$$\begin{vmatrix} a & b \\ a_1 & b_1 \end{vmatrix} \neq 0,$$

then u and v are linearly expressible in terms of x and y with rational coefficients, that is, the transformation is birational. The following is a more complicated example. Let L be the curve

$$\begin{aligned} y^2 &= x^4 - x^3 + 2x - 2 \\ &= (x-1)(x^3+2). \end{aligned} \qquad (*)$$

We will show that L can be birationally transformed into a curve L' of the form $v^2 = \varphi_3(u)$, where $\varphi_3(u)$ is a cubic polynomial. To this end we divide both sides of the equation $(*)$ by $(x-1)^4$ and put

$$x - 1 = \frac{1}{u}, \qquad \frac{y}{(x-1)^2} = v.$$

Then equation $(*)$ becomes

$$v^2 = 3u^3 + 3u^2 + 3u + 1.$$

Also, x and y are rationally expressible in terms of u and v,

$$x = \frac{1+u}{u}, \qquad y = \frac{v}{u^2},$$

and conversely,

$$u = \frac{1}{x-1}, \qquad v = \frac{y}{(x-1)^2},$$

that is, the curves L and L' are birationally equivalent [3].

With the possible exception of finitely many points, it is possible to establish a one-to-one correspondence between the sets $\mathcal{M}$ and $\mathcal{M}'$ of two birationally equivalent curves. The exceptional points are the points for which both the numerator and denominator of at least one of the functions in (3) or (3′) vanish (this implies that there are no exceptional points in the case of linear transformations). In our second example, both the numerator and denominator of the function

$$v = \frac{y}{(x-1)^2}$$

vanish at the point $(1,0)$ on L. This means that no point on L' corresponds to the point $(1,0)$ on L.

From the viewpoint of diophantine analysis equivalent curves have "the same standing." Their orders need not be the same. But it can be shown that two birationally equivalent curves have the same genus. In other words, while neither the order n nor the number d of double points is invariant under a birational transformation, the genus p of a curve, which is a function of these quantities, is such an invariant.

The converse assertion is false: curves with the same genus need not be birationally equivalent. This is clear from the earlier example of the curves

$$x^2 + y^2 = 1 \quad \text{and} \quad x^2 + y^2 = 3.$$

Both of these curves have genus 0, but the first has infinitely many rational points and the second none.

Thus curves of the same genus split into equivalence classes of birationally equivalent curves. The full significance of these concepts came to light in the works of Henri Poincaré who, at the beginning of this century, made the class of birational transformations the very basis of the classification and investigation of the problems of diophantine analysis. We will discuss these matters in Chapter 12.

At this time we note a fact that is of great importance for what follows: if Γ is a cubic curve with at least one rational point, then it is possible to reduce its equation by means of birational transformations to an equation of the form

$$y^2 = x^3 + ax^2 + b, \tag{5}$$

where a and b are rational. In what follows we will often assume that the equation of Γ is given in the form (5).

Notes

[1] A field is a system of elements closed under the four arithmetical operations (with the exception of division by zero) . Examples of fields are the rational numbers, the numbers of the form $a + b\sqrt{2}$ with a and

b rational, and the real numbers. Unless otherwise specified, it is safe to assume that the field involved is the field Q of rational numbers.

[2] It is clear that the sum and difference of numbers of the form $a+b\sqrt{3}$ is again of this form. The reader can check that the quotient of two such numbers is again a number of this form, that is, that $Q(\sqrt{3})$ is a field.

[3] It is easy to see that a polynomial of the form $y^2 = f_{2n}(x)$, where $f_{2n}(x)$ is a polynomial of degree $2n$ with rational root a, that is,

$$f_{2n}(x) = (x - a)\, g_{2n-1}(x),$$

can be transformed into a curve

$$v^2 = \varphi_{2n-1}(u).$$

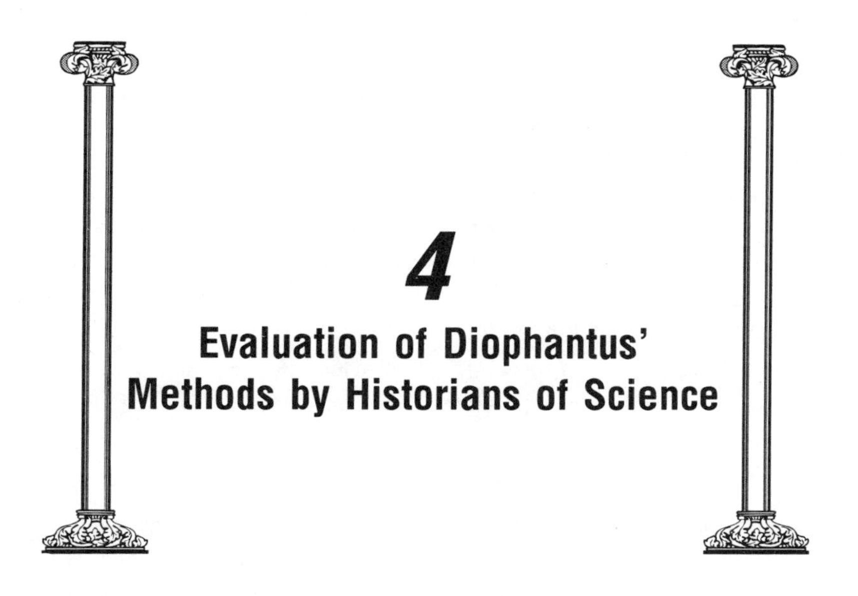

4
Evaluation of Diophantus' Methods by Historians of Science

In what follows we will show that Diophantus had a general method for the determination of rational points on quadratic curves. Poincaré showed that this method is applicable to all curves of genus 0 with a rational point. Diophantus also found general methods for obtaining rational points on cubic curves which differed in a fundamental way from those he used for quadratic curves. Poincaré's papers show that these methods of Diophantus can be used to obtain rational points on arbitrary curves of genus 1. To this day there are no other general methods for finding rational points on algebraic curves.

We will also show what role the ideas and methods of Diophantus played in the history of mathematics and how mathematicians from Viète and Fermat to Euler used them.

To this day, the majority of historians of science, in contradistinction to mathematicians, have underestimated the works of Diophantus. Many of them thought that Diophantus limited himself to finding a single solution, and to obtain it used different tricks for different problems. For example, H. Hankel wrote:

> ... having mastered 100 of Diophantus' solutions, a modern mathematician would find it difficult to solve the 101st problem ... Diophantus dazzles rather than delights

(H. Hankel, Zur Geschichte der Mathematik im Altertum und Mittelalter, Leipzig, 1874, p. 165).

One might try to argue that such an evaluation is due to the fact that Hankel's book was published before the appearance of Poincaré's papers which shed a new light on the issue of diophantine equations. But this is hardly persuasive in view of, say, the 1951 *History of Mathematics* by O. Becker and J. Hofmann (O. Becker, J. Hofmann, Geschichte der Mathematik, Bonn). On p. 90 we read:

> Diophantus gives no general method whatsoever but seems to be applying for each new problem a new unexpected artifice reminiscent of the East.

Van der Waerden makes similar remarks in his "Science Awakening":

> He (that is Diophantus) is usually satisfied when he has found one solution; it makes no difference to him whether the solution is integral or fractional. His method varies from case to case. ([6], p. 279.)

In connection with indeterminate quadratic equations he says that

> (Diophantus) knows how to arrange matters cleverly so that either the term in x^2 drops out, or the constant term, so that he can solve for x rationally. ([6], p. 283.)

Similarly, the 1961 edition of "Mathematisches Woerterbuch" states (under the heading Diophantus of Alexandria) that

> The solution of indeterminate equations of second and higher degrees is achieved ... through the use of artifices which vary from case to case. (Quote due to L. Boll)

H.G. Zeuthen's evaluation of Diophantus is more balanced:

> In general, Diophantus tries to find some one solution of a problem, without looking for its general solution which includes all possible particular solutions. But if one wants to understand Diophantus' results, then one should not attach special importance to this fact, for what is special about his solution is that he immediately assigns specific values to the auxiliary quantities which serve to solve the problem.

([7]. H.G. Zeuthen, Geschichte der Mathematik im Altertum und Mittelalter, Copenhagen 1886. Neudruck 1966.) Elsewhere he states that

> In case of a given but arbitrary number, (Diophantus) always chooses a definite number and calculates with it. But his calculations are so transparent that they yield in effect a general method. When it comes to unknowns, he often begins by testing certain values which, in general, fail to do what was required, namely satisfy a proposed equation—say, make a certain expression a square, and so on. But he remembers the calculations so well that he can immediately think of the changes needed in the selected number. ([7]. Quote due to L. Boll.)

Zeuthen goes on to analyze Diophantus' methods for the solution of indeterminate quadratic equations. But he too fails to see in Diophantus' work methods for the solution of indeterminate cubic equations. To this day various mathematicians of the new era are credited with these methods. Thus in his "Diophantine Equations" (T. Skolem, Diophantische Gleichungen, Berlin, 1938) Skolem credits Cauchy and Lucas with Diophantus' methods, and Lucas—Cauchy and Fermat. Clearly, when it comes to credit, Diophantus seems to have been just as unlucky with general methods for the solution of indeterminate equations as with negative numbers.

But now let us consider his equations.

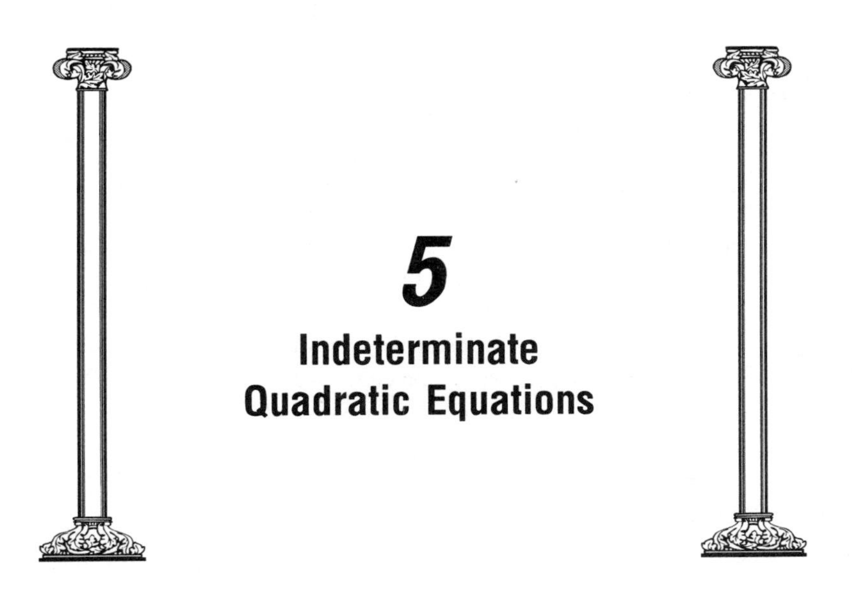

5 Indeterminate Quadratic Equations

Before Diophantus, two types of quadratic equations had been considered, namely

$$x^2 + y^2 = z^2 \quad \text{and} \quad x^2 - ay^2 = 1.$$

The first of these was dealt with in ancient Babylonia. The formulas for its solution,

$$x = k^2 - 1, \quad y = 2k, \quad z = k^2 + 1,$$

were found by the Pythagoreans. For $a = 2$ all integral (rather than rational) solutions of the second of these equations are found in Euclid's "Elements." Archimedes, who put before Eratosthenes the well-known "cattle problem," probably knew its solution for an arbitrary nonsquare a.

In book II of his "Arithmetic" Diophantus considered various indeterminate second-order equations and essentially established the following result: *An indeterminate second-order equation in two unknowns has either no rational solutions or infinitely many. In the latter case, all the solutions are expressible as rational functions of a single parameter*

$$x = \varphi(k), \quad y = \psi(k),$$

where φ and ψ are rational functions.

To show this we first quote problem 8 in book II:

> To divide a given square into a sum of two squares.
>
> To divide 16 into a sum of two squares.
>
> Let the first summand be x^2, and thus the second $16 - x^2$. The latter is to be a square. I form the square of the difference of an arbitrary multiple of x diminished by the root [of] 16, that is, diminished by 4. I form, for example, the square of $2x - 4$. It is $4x^2 + 16 - 16x$. I put this expression equal to $16 - x^2$. I add to both sides $x^2 + 16$ and subtract 16. In this way I obtain $5x^2 = 16x$, hence $x = 16/5$.
>
> Thus one number is $256/25$ and the other $144/25$. The sum of these numbers is 16 and each summand is a square.

We will now try to state Diophantus' method "in pure form." Thus consider the equation

$$x^2 + y^2 = a^2, \tag{6}$$

which represents a circle with center at the origin. One of its rational solutions is $(0, -a)$. Diophantus makes the substitution

$$x = x, \quad y = kx - a. \tag{7}$$

Since he has no symbol for an arbitrary k, he takes $k = 2$, but notes that one should form " the square of the difference of an arbitrary multiple of x diminished by the root [of] 16," that is, in our symbolism, the square of $kx - 4$.

In geometric terms, the substitution (7) amounts to drawing the straight line

$$y = kx - a \tag{7'}$$

through $(0, -a)$. This straight line intersects the circle (6) in a second point whose coordinates are rational functions of k. Indeed,

$$x^2 + (kx - a)^2 = a^2$$

and

$$x = \frac{2ak}{k^2+1}, \qquad y = kx - a = a\frac{k^2-1}{k^2+1}.$$

Thus to every rational value of k there corresponds just one rational point on the curve (6). It is easy to see that, conversely, if we join any rational point on the curve (6) to $(0, -a)$, then we obtain a straight line with a rational slope.

The solution of problem 9 in book II makes Diophantus' method even clearer. He states this problem as follows:

> To divide a given number which is the sum of two squares into two other squares.

Diophantus gives the number 13, which is equal to the sum $4+9$. Thus one solution, namely $(2, 3)$, is already known. To find a second solution, Diophantus takes as the first number $x = t+2$ and as the second number $y = 2t - 3$. In other words, he draws a straight line through the point $(2, -3)$ and notes, as before, that instead of the multiplier 2 one can choose any other number.

It is worth noting that Diophantus takes as the known point not the point with positive coordinates, which we gave, but a point with a negative coordinate. This corresponds to a negative solution. In general, in intermediate computations Diophantus gladly operates with negative numbers, although each final solution is always rational and positive.

Diophantus applies the same procedure in problems 16, 17, and in other problems in book II.

It is easy to see that Diophantus' method is quite general, and that it makes it possible to find all rational points on a quadratic curve with at least one rational point. Indeed, consider a quadratic curve

$$f_2(x, y) = 0 \tag{8}$$

in two variables with a rational point (a, b). Following Diophantus, we make the substitution

$$x = a + t, \quad y = b + kt$$

and obtain

$$f_2(a+t, b+kt) = f_2(a,b) + tA(a,b) + ktB(a,b) + t^2C(a,b,k) \\ = 0.$$

Since $f_2(a,b) = 0$, it follows that

$$t = -\frac{A(a,b) + kB(a,b)}{C(a,b,k)}.$$

Thus for every rational k we obtain just one rational solution.

If the given equation is of the form

$$y^2 = a^2x^2 + bx + c, \tag{9}$$

then Diophantus makes a slight change in his method and puts

$$y = ax + m.$$

Then

$$x = \frac{c - m^2}{2am - b}.$$

We will try to explain the geometric sense of this substitution. To do this we will go over to homogeneous, or projective, coordinates. Such coordinates are very convenient for the investigation of the properties of algebraic curves and we will frequently use them in the future. This being so, we will consider this issue in some detail.

Thus far, following the common approach of analytic geometry, we have considered the affine plane $R^{(2)}$, each of whose points is given by an ordered pair (x, y) of real numbers. Now we will consider the projective plane $P^{(2)}$, each of whose points is given by an ordered triple (u, v, z) of real numbers not all of which are zero. We identify points (u, v, z) and (u_1, v_1, z_1) if and only if $u_1 = ku$, $v_1 = kv$, and $z_1 = kz$ for $k \neq 0$. Thus infinitely many triples determine the same point. The numbers of an arbitrary triple which determines a point are called its *homogeneous coordinates*.

We will now determine a (partial) one-to-one correspondence between the points of the planes $R^{(2)}$ and $P^{(2)}$. Let (u, v, z) be a point on $P^{(2)}$. If $z \neq 0$, then instead of (u, v, z) we take the triple $(u/z, v/z, 1)$

(which determines the same point on $P^{(2)}$) and associate with it the point (x, y) on $R^{(2)}$ with $x = u/z$, $y = v/z$.

If $z = 0$, then there is no point on $R^{(2)}$ which corresponds to the point $(u, v, 0)$ on $P^{(2)}$. We will call such points *points at infinity.* All such points lie on the *line at infinity* given by $z = 0$. Since the z-coordinate is just as "good" as the other coordinates, we have every right to regard the points at infinity and the line at infinity on $P^{(2)}$ as no less "good" than its finite points and lines.

In order to go over from an equation

$$f(x, y) = 0$$

in affine coordinates to one in homogeneous coordinates, we put

$$x = u/z, \quad y = v/z,$$

and end up with an equation of the form

$$\Phi(u, v, z) = 0,$$

where $\Phi(u, v, z)$ is a polynomial in u, v, z. For example, the equation of the hyperbola

$$x^2 - y^2 = 1,$$

rewritten in homogeneous coordinates, is

$$u^2 - v^2 = z^2.$$

To find its points at infinity, that is, its points of intersection with the line at infinity, we put $z = 0$. Then $v = \pm u$. In other words, our hyperbola has two points at infinity given by $(1, 1, 0)$ and $(1, -1, 0)$ respectively[1]. Both of these points have rational coordinates. Such points are called *rational points at infinity.*

We now turn to Diophantus' substitution. Using homogeneous coordinates we can rewrite equation (9) as

$$v^2 = a^2u^2 + buz + cz^2. \tag{9'}$$

Its rational points at infinity are $(1, a, 0)$ and $(1, -a, 0)$. We draw a straight line through the first of these points. The general equation of a line in homogeneous coordinates is

$$Au + Bv + Cz = 0.$$

But $(1, a, 0)$ is on this line, that is,

$$A \cdot 1 + B \cdot a + C \cdot 0 = 0.$$

We can therefore put $A = ka$, $B = -k$, $C = km$, where m is arbitrary. Hence the equation of our straight line is

$$au - v + mz = 0,$$

or, in terms of affine coordinates,

$$y = ax + m.$$

But this is the substitution used by Diophantus. As such, it is equivalent to drawing an arbitrary straight line through a rational point at infinity on the curve (9).

At this point we stress that we do not claim that Diophantus knew of points at infinity on curves. He simply used equivalent arguments. In the history of mathematics there are many instances when the basic facts of a theory were discovered prior to the emergence of that theory and of its fundamental concepts. For example, this happened in the case of the arithmetic of quadratic fields. This arithmetic was constructed by Euler, Lagrange, and Gauss not only before the introduction of quadratic fields but also before the introduction of the concept of an algebraic number. The whole development transpired within the framework of the theory of quadratic forms, but the facts discovered in this theory were equivalent to the arithmetic of quadratic fields.

The same sort of thing happened in the case of Diophantus' "Arithmetic." Here certain considerations of algebraic geometry were discovered and investigated within the framework of pure algebra and number theory, without any geometric interpretations.

We now consider the following question: Was Diophantus aware that his problems had infinitely many solutions or was he actually satisfied with finding a single rational solution?

In book II Diophantus says nothing about this matter, and the infinity of solutions can only be inferred from his method. But in problem 19 in book III he writes "and we have learned how we can represent a square as a sum of squares in infinitely many ways."

Finally, in two lemmas associated with problems in book VI, Diophantus proves the following: If the indeterminate equation

$$ax^2 + b = y^2$$

has a rational solution (x_0, y_0), then it has infinitely many such solutions. The first lemma pertains to problem 12 and states that

> When two numbers are given whose sum is a square, then squares are found in infinitely many ways such that each of them, when we multiply it by one of the given numbers and add the second, yields a square.

In other words, if in the above equation b is positive and $a + b$ is a square, then the equation has infinitely many solutions.

But what is the significance of the condition that $a + b$ is a square? This is not difficult to see. If $a + b = m^2$, then our equation has the rational solution $(1, m)$. To prove his assertion, Diophantus puts

$$x = t + 1, \quad y = y,$$

and obtains the equation

$$at^2 + 2at + m^2 = y^2,$$

whose constant term is a square. This being so, he can find the remaining rational solutions by means of the usual method, that is, by putting

$$y = kt - m.$$

Then he obtains

$$t = 2\frac{a + km}{k^2 - a}.$$

Now the unknowns x and y can be expressed as rational functions of a single parameter.

Throughout his argument Diophantus uses the particular values $a = 3$, $b = 6$, that is, $m^2 = 9$, but his method of proof is general.

It is interesting that 1500 years later Euler (for more about his work see Chapter 10) used the same substitution in his "Algebra." He assumes that the equation

$$y^2 = Ax^2 + Bx + C$$

has a rational solution (x_0, y_0). He puts $x = t + x_0$, $y = y$ and obtains the new equation

$$pt^2 + qt + r = y^2.$$

Its constant term is easily seen to be y_0^2. Now he makes the usual diophantine substitution.

The second lemma pertains to problem 15 in book VI and is of a far more general character

> Given two numbers. If subtraction of one of them from the product of a square by the other yields a square, then it is possible to find another square, greater than the first, that yields a similar outcome.

In other words, if the equation

$$ax^2 - b = y^2 \quad (b > 0)$$

has a rational solution (p, q), then it has also a has a larger solution (p_1, q_1). In turn, starting with (p_1, q_1), one can obtain a larger solution (p_2, q_2). Here "larger" means that $p < p_1 < p_2 < \cdots$ and $q < q_1 < q_2 < \cdots$.

Diophantus carries out the proof for the case $a = 3$, $b = 11$. His first solution is $(5, 8)$. Using the substitution $x = t + p$ ($p = 5$), he obtains the equation

$$at^2 + 2apt + q^2 = y^2,$$

which he solves by the method of the previous lemma.

In both lemmas, Diophantus' method of proof, while illustrated by means of an example, is completely general.

We see that Diophantus not only *discovered* the theorem stated at the beginning of this chapter but also *proved it in full generality.*

We note that Diophantus' method for the solution of indeterminate equations of the form

$$y^2 = ax^2 + bx + c$$

coincides with the so-called "Euler substitution," familiar to students of mathematical analysis. In both cases x and y are expressed in terms of rational functions of a single parameter, and in both cases this is done by means of the same substitutions. The only difference is that when computing the integral

$$\int \frac{dx}{\sqrt{ax^2 + bx + c}}$$

we need not require the coefficients of the functions involved to be rational, and can therefore put

$$y = \sqrt{x} + t, \quad \text{or} \quad y = xt + \sqrt{c}.$$

Not so in Diophantus' case. Here the points involved are rational, and therefore the coefficients in all the substitutions had to be rational. This was the additional requirement which Diophantus had to take into consideration.

Notes

[1] Since $v = u$ or $v = -u$, it follows that the points in question are $(u, u, 0)$ and $(u, -u, 0)$. Multiplication by $1/u$ yields $(1, 1, 0)$ and $(1, -1, 0)$.

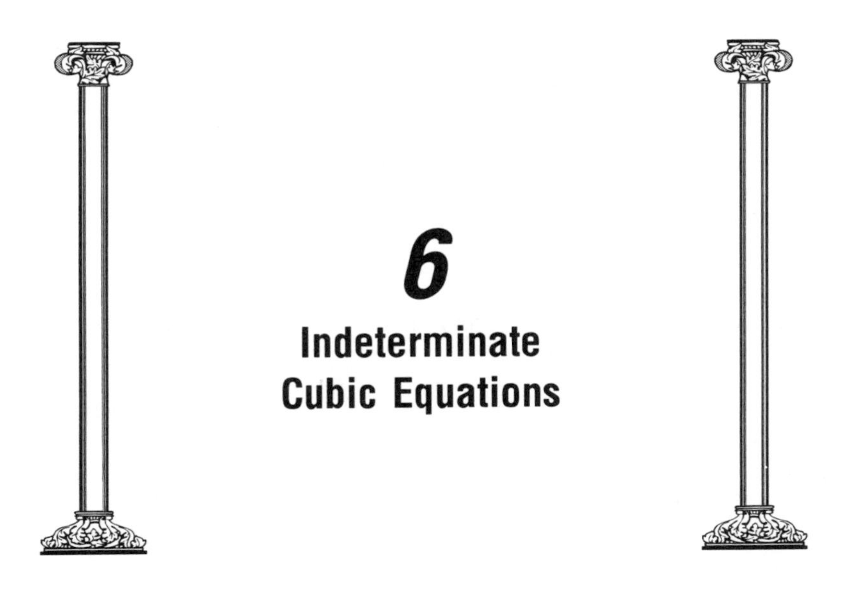

6 Indeterminate Cubic Equations

In book IV Diophantus considers cubic and quartic indeterminate equations. Here things are far more involved. Even if a cubic does have rational points, their coordinates cannot, in general, be expressed in terms of rational functions of a single parameter. However, if we know one or two rational points on a cubic curve, then we can find an additional rational point on it. Indeed, an arbitrary straight line intersects a cubic curve in three points whose coordinates can be determined from, say, the cubic equation obtained by the elimination of y from the equations of the curve Γ,

$$f_3(x, y) = 0, \qquad (10)$$

and of the straight line. If two of the roots of the resulting equation are rational, then so is the third. (To see this, note, for example, that the sum of the roots of a cubic equation is the quotient of the negative of the coefficient of x^2 divided by the coefficient of x^3. If the coefficients and two of the roots of the equation are rational, then so is the third root.) This observation underlies the following two procedures:

1. If P is a rational point on the curve Γ, then we draw at P the tangent to Γ with rational slope k which will intersect Γ at an additional

rational point. (Indeed, by solving simultaneously the equations of the curve and the tangent we obtain a cubic equation with a double rational root. This means that its third root is also rational.)

2. If P_1 and P_2 are rational points on Γ, then the straight line P_1P_2 intersects Γ at an additional rational point.

We will refer to these two procedures as *the tangent and secant methods of Diophantus.* To justify this name we return to his problems. Problem 24 in book IV:

> To divide a given number into two numbers such that their product is a cube minus its side.
>
> Given the number 6. I take the first number to be x. Then the second will be $6 - x$.
>
> The condition to be satisfied is $x_1x_2 = y^3 - y$. But $x_1x_2 = 6x - x^2$. This expression is to be equal to $y^3 - y$. Thus I form $y = ax - 1$, where a is arbitrary, for example, $a = 2$. So I now form
>
> $$(2x - 1)^3 - (2x - 1) = 8x^3 - 12x^2 + 4x.$$
>
> This expression is to be $6x - x^2$. If the coefficients of x in both expressions were equal, x would be rational. 4 arises out of $3\cdot 2 - 2$, but 6 comes from the data. I must therefore determine a so that $3a - a$ is equal to 6. I must therefore put $y = 3x - 1$ and I obtain
>
> $$y^3 - y = 27x^3 - 27x^2 + 6x.$$
>
> This expression must be equal to $6x - x^2$. It follows that $x = 26/27$. Therefore
>
> $$x_1 = 26/27, \quad x_2 = 136/27.$$

We will now try to set down Diophantus' method "in pure form." Let a denote the given number and x and $a - x$ the required numbers. We know that

$$x(a - x) = y^3 - y. \tag{11}$$

One of the rational solutions is $(0,-1)$. Following Diophantus, we pass through this point the line

$$y = kx - 1 \tag{$*$}$$

(Diophantus takes initially $k = 2$) and find its point of intersection with the curve (11):

$$ax - x^2 = k^3x^3 - 3k^2x^2 + 2kx.$$

For x to be rational it suffices to put

$$2k = a, \quad \text{that is,} \quad k = a/2, \tag{$**$}$$

which is what Diophantus does. Then we obtain

$$x = \frac{3k^2 - 1}{k^3} = 2\frac{3a^2 - 4}{a^3}.$$

To explain the significance of the condition $(**)$ for the straight line $(*)$ we apply Diophantus' method to the arbitrary cubic equation in two variables (10) with a rational solution $(a,b) : f_3(a,b) = 0$. We draw through $P(a,b)$ the straight line

$$y - b = k(x - a), \tag{12}$$

or

$$x = a + t, \quad y = b + kt. \tag{13}$$

Then

$$\begin{aligned} f_3(a+t, b+kt) &= f_3(a,b) + tA(a,b) + ktB(a,b) \\ &\qquad + t^2C(a,b,k) + t^3D(a,b,k) \\ &= 0. \end{aligned}$$

But $f_3(a,b) = 0$ and, if we put

$$A(a,b) + kB(a,b) = 0, \tag{14}$$

then we obtain

$$k = -\frac{A(a,b)}{B(a,b)} = -\left(\frac{\partial f_3}{\partial x} \Big/ \frac{\partial f_3}{\partial y}\right)(P),$$

that is, the slope of the line (12) must be chosen so that it is tangent to the curve (10) at the point $P(a, b)$. Thus here Diophantus uses the method of tangents.

Diophantus uses the same method to solve problem 18 in book VI, as well as, very likely, the problem

$$x^3 + y^3 = a^3 - b^3.$$

According to him, he solved the latter problem in his book "Porisms," which has not come down to us.

We note that in the course of his computations Diophantus obtained a purely computational method for the determination of the slope k of the tangent $\frac{dy}{dx}$, or

$$\frac{dy}{dx} = -\frac{\partial f_3}{\partial x} \Big/ \frac{\partial f_3}{\partial y}.$$

This method, which dispenses with the need for limits and can therefore be implemented purely algebraically (in a field without a topology), played an important role in the historical process of formation of the derivative—especially in the work of Fermat and Descartes—and is now widely used in algebraic geometry.

We now go over to problem 26 in book IV in which Diophantus uses the method of secants.

> To find two numbers such that their product augmented by either gives a cube.
>
> I put $x_1 = a^3 x$ with $a = 2$, say, so that $x_1 = 8x$. Let $x_2 = x^2 - 1$. Then one condition is satisfied, for $x_1 x_2 + x_1$ is a cube.
>
> It remains to fulfill the requirement that $x_1 x_2 + x_2$ is also a cube. But $x_1 x_2 + x_2 = 8x^3 + x^2 - 8x - 1$. I put this expression equal to the cube of $2x - 1$, that is, equal to $8x^3 - 12x^2 + 6x - 1$. Then $x = 14/13$.
>
> But then $x_1 = 112/13$, $x_2 = 27/169$.

Following Diophantus, we denote the first unknown by $a^3 x$ and the second by $x^2 - 1$. Then the first condition of the problem is satisfied and the second yields

$$a^3 x^3 + x^2 - a^3 x - 1 = y^3. \tag{15}$$

Diophantus makes the substitution $y = ax - 1$ and obtains

$$x = \frac{a^3 + 3a}{1 + 3a^2}.$$

We consider in some detail the method used by Diophantus in this case. One of the rational solutions of (15) is $(0, -1)$. We draw the straight line $y = kx - 1$ through this point and find its point of intersection with (15):

$$(a^3 - k^3)x^3 + (1 + 3k^2)x^2 - (a^3 + 3k)x = 0.$$

In the previous case, Diophantus put the coefficient of x equal to zero. Here he puts the coefficient of x^3 equal to zero and obtains

$$a^3 - k^3 = 0, \quad k = a.$$

What is the geometric significance of this step? To answer this question we rewrite (15) in terms of homogeneous coordinates. Putting $x = u/z$, $y = v/z$ we obtain

$$a^3u^3 + u^2z - a^3uz^2 - z^3 = v^3. \tag{15'}$$

We see that this curve has the rational points $P_1(0, -1, 1)$ and $P_2(1, a, 0)$ which determine the line

$$v = au - z.$$

The intersection of (15′) and this line yields a third rational point. Thus, in this case, Diophantus uses the method of secants when one of the rational points is finite and the other is a point at infinity.

Diophantus also uses his methods of tangents and secants in other problems in books IV and VI.

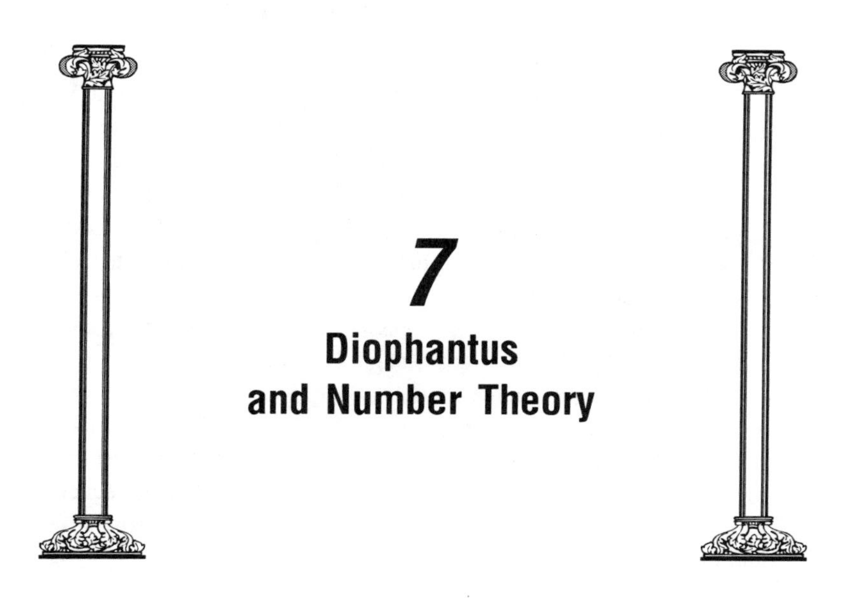

7
Diophantus and Number Theory

The books of the "Arithmetic" which have come down to us contain no investigations in number theory in the strict sense of the term. However, when stating a problem or solving it, Diophantus sometimes includes the conditions under which it is solvable or unsolvable, or notes that a number obtained in the process of solution cannot be represented as, say, a sum of two squares. This is how theorems of number theory turn up in the "Arithmetic." To judge by a remark of Diophantus, he considered these and similar theorems in a special book titled "Porisms," which has not come down to us. This being so, the best we can do to form an opinion about Diophantus' knowledge of number theory is to rely on the remarks and diorisms (restrictions) found in the "Arithmetic." We begin with problem 19 in book III:

> To find four numbers such that the square of their sum plus or minus any one singly gives a square.
>
> Since in any right triangle, if we add to the square on the hypotenuse, or subtract from it, twice the product of the sides of the right angle we obtain a square, I look first for four right triangles with the same hypotenuse. This is the same as dividing a square into a sum of two squares in four ways, and we know infinitely many ways of dividing a square into a sum of two squares.

Take two right triangles in the smallest numbers 3, 4, 5 and 5, 12, 13. Multiply the first by the hypotenuse of the second and vice versa. Then the first triangle will be 39, 52, 65 and the second 25, 60, 65. These are right triangles with equal hypotenuses.

65 can be divided into a sum of two squares in two ways, into 16 and 49 and into 64 and 1. This is because 65 is the product of 13 and 5, each of which numbers is the sum of two squares.

Now from 49 and 16 I take roots, 7 and 4, and form from the two numbers 7 and 4 the right triangle 33, 56, 65.

Similarly, 64 and 1 have roots 8 and 1, and I form from them another right triangle whose sides are 16, 63, 65. Thus we have obtained four right triangles with equal hypotenuses. Now I return to the original problem. I put the sum of the four numbers equal to $65x$, form the fourfold area of each triangle, multiply it by x^2, and obtain in this way

$$x_1 = 4056x^2, \quad x_2 = 3000x^2, \quad x_3 = 3696x^2, \quad x_4 = 2016x^2.$$

The sum of these four numbers is $12768x^2$ and is equal to $65x$. This gives $x = 65/12768$. Thus

$$x_1 = 17136600 \cdot 1/n,$$

$$x_2 = 12675000 \cdot 1/n,$$

$$x_3 = 15615600 \cdot 1/n,$$

$$x_4 = 8517600 \cdot 1/n,$$

with $n = 163021824$.

This problem is remarkable in many respects. Here Diophantus talks for the first time about triangles "in the smallest numbers" and about forming such triangles from "two numbers." Of course, the issue here is finding rational solutions of the indeterminate equation

$$x^2 + y^2 = z^2,$$

which we discussed in Chapter 5. The most general formulas for its solution were given by Euclid in his "Elements." Diophantus uses them without special stipulations. These formulas are:

$$z = p^2 + q^2,$$
$$x = 2pq,$$
$$y = p^2 - q^2.$$

They yield all primitive solutions of the equation for coprime p and q with different parities. (Since the equation is homogeneous, the extension of the domain of solutions from the integers to the rationals yields nothing new.) These solutions can be obtained by the same method which Diophantus used in problem 8 of book II to divide a given square into a sum of two squares (see Chapter 5).

Moreover, this problem contains the assertion that the product of two integers each of which is a sum of two squares is itself representable as a sum of two squares in at least two ways (provided that the two integers are unequal). Specifically, if

$$p = a^2 + b^2 \quad \text{and} \quad q = c^2 + d^2,$$

then

$$\begin{aligned} pq &= (ac + bd)^2 + (ad - bc)^2 \\ &= (ad + bc)^2 + (ac - bd)^2. \end{aligned}$$

It was in remarks related to this problem that Fermat stated his famous assertion that every prime of the form $4n + 1$ is representable as a sum of two squares in just one way. In the same connection he gave a method for determining in how many ways a given number can be written as a sum of two squares.

Did Diophantus know these propositions? To answer this question we consider another problem, supplied with diorisms (restrictions) which deal with the representation of a number as the sum of two squares. This is problem 9 in book V:

> To divide unity into two parts such that, if the same number be added to either part, the result will be a square.

The statement is followed by a diorism which must be imposed on the given number for the problem to be solvable. Unfortunately, the text following the words "The given number must not be odd, and double the number plus unity..." is distorted. There are reconstructions which we will consider below. But first the text of the problem.

> We are to add to each part 6 so that the result be squares.
>
> Since we want to divide 1 so that upon addition of 6 to each part we get a square, the sum of the squares will be 13. Thus 13 must be divided into two squares each of which is greater than 6.
>
> If I divide 13 into two squares whose difference is less than 1, then the problem is solved. I take half of 13, obtain $6\frac{1}{2}$, and ask what fraction increased by $6\frac{1}{2}$ is a square. I multiply by 4. Thus I look for a square fraction which when added to 26 yields a square. Thus $26 + \frac{1}{x^2}$ is a square, hence also $26x^2 + 1$. I put
>
> $$26x^2 + 1 = (5x + 1)^2$$
>
> and obtain $x = 10$. Thus $x^2 = 100$, $1/x^2 = 1/100$.
>
> This means that what is added to 26 is 1/100, which means that what is added to $6\frac{1}{2}$ is $1/400$, which gives the square $(51/20)^2$.
>
> So it is necessary to take the root of each of the squares, whose sum is 13, close to $51/20$. And so I ask what number subtracted from 3 and added to 2 will give that much, namely $51/20$.
>
> So I form the squares of $11x + 2$ and $3 - 9x$. Their sum is to be 13. Hence
>
> $$202x^2 - 10x + 13 = 13,$$
>
> and so $x = 5/101$.
>
> This means that the root of one square will be 257/101 and of the other 258/101. And if from the squares of each I remove 6, then I get as parts of unity 5358/10201 and 4843/10201, and it is clear that each together with 6 yields a square.

We wish to point out that when solving this problem Diophantus made use of interesting new methods that historians of mathematics have paid little attention to. The equation $ax^2 + 1 = y^2$ $(a = 26)$ with which he is concerned is now known as the *Fermat equation.* It was of great interest in the 17th and 18th centuries. In subsequent problems Diophantus refers to the procedure of finding squares with given sum each of which must satisfy an inequality as an *approximation procedure*. Basically, the problem just presented deals with approximating $\sqrt{26}$. We will not explore the many issues that arise in connection with this problem and will focus on the diophantine restrictions. But we suggest that the reader should work through the problem and try to interpret it geometrically.

[Translator's note: This paragraph was taken from the German translation of this book published by VEB Deutscher Verlag der Wissensschaften, Berlin, 1974.]

We can write the conditions of the problem in the form of the system of equations

$$x + y = 1,$$

$$x + a = u^2,$$

$$y + a = v^2.$$

Addition of the last two of these equations yields

$$2a + 1 = u^2 + v^2.$$

Therefore a must be chosen so that $2a + 1$ is a sum of two squares.

It can be shown (see [5], p.270) that a number is representable as a sum of two squares if and only if its squarefree part is not divisible by a prime of the form $4n - 1$. How close was Diophantus' diorism to this condition? Since we have only its distorted version (see p. 36), the answer to this question must be based on a reconstruction of the text.

One of the famous mathematicians of the 19th century Carl Gustav Jacob Jacobi (1804–1851), a younger contemporary of Gauss, devoted a special investigation to this issue ("Ueber die Kenntnisse des Diophantus von der Zusammensetzung der Zahlen." Berliner Monatsberichte 1847 (Gesammelte Werke, VII, 1891, p. 336)). He carried out a detailed philological investigation of Diophantus' text and proposed the following reconstruction of it:

"The given number (that is a) must not be odd and double the number plus 1 must not be divisible by any number which, when increased by 1, is divisible by 4."

A similar reconstruction of this text was subsequently made by Paul Tannery, the great expert on antiquity and editor of Diophantus (the 1893 edition).

This condition is necessary only if we add to it the stipulation "after division by the largest square in it." It seems that Diophantus assumed it. But then the condition is also sufficient, that is, it fully characterizes the set of whole numbers representable as a sum of two squares.

Jacobi assumed that Diophantus had a proof that his condition was necessary, that is, that he could justify his diorism. In his paper he gives a reconstruction of such a proof using only methods used by Euclid and Diophantus in their works.

Jacobi was sure that Diophantus knew that his condition was also sufficient but could not prove this because a proof required means beyond those of ancient mathematics.

We note that after Diophantus it was only Fermat who gave a general condition for a number not to be representable as a sum of two integral or fractional squares. Here is Fermat's formulation:

> if an integer has a prime divisor of the form $4n - 1$ and has no square divisor, then it cannot be a sum of two squares ("not even in fractions") [19], p. 63.

This "negative" criterion is equivalent to the "positive" criterion for the representability of a number as a sum of two squares stated on p.29. Both can be deduced from a remarkable theorem formulated by Fermat and proved by Euler. It states that the primes representable as sums of two squares are precisely those of the form $4n + 1$.

In problem 14 of book V Diophantus states a necessary condition for a number to be representable as a sum of three squares. The diorism is to the effect that the number must not be of the form $8n + 7$. Here the proof of necessity probably presented no difficulty for Diophantus, but nowhere does he assert that every odd number not of the form $8n + 7$ is representable as a sum of three squares; this in spite of the fact that he could have found this proposition in a purely inductive manner.

Bearing in mind the principle of the ancient mathematicians that one only stated propositions which one could prove, it is safe to assert that Diophantus could prove all his diorisms. But then he was not only a brilliant algebraist, not only the founder of diophantine analysis, but also an outstanding number theorist.

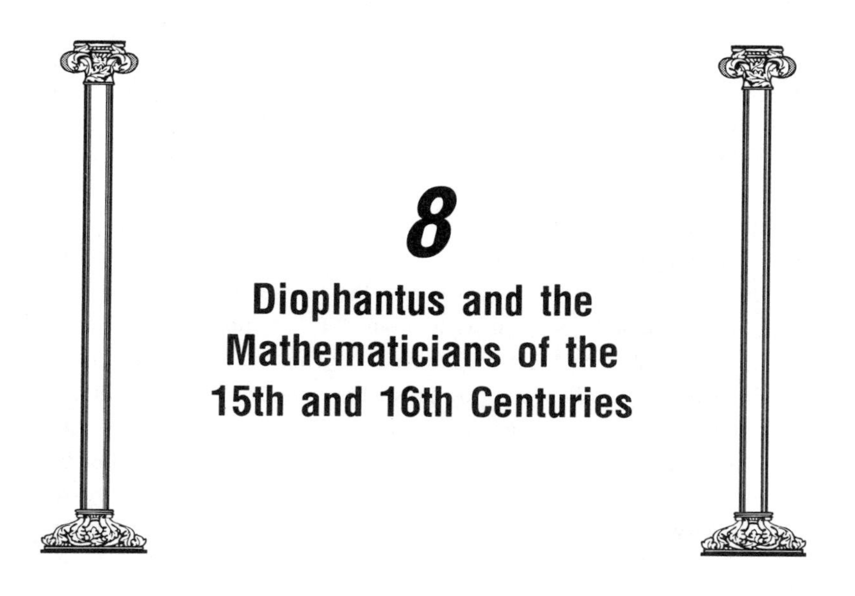

8 Diophantus and the Mathematicians of the 15th and 16th Centuries

There were commentators on the work of Diophantus in antiquity. The works of the famous Hypatia, daughter of the Alexandrian scholar Theon, were devoted to an analysis of his books. Hypatia lived at the end of the 4th and at the beginning of the 5th centuries. She acquired fame as a brilliant orator and expert on Plato's philosophy. Unfortunately, her works have not come down to us.

We know of no Alexandrian mathematician after Hypatia. The last Greek scholars, Proclus, Isidorus, and Simplicius, worked in Athens. But here too scientific thought was extinguished at the beginning of the 7th century. Antique science perished together with antique society. In the period between the 9th and 13th centuries there appeared new scientific centers in Constantinople, Baghdad, and other cities of the Arabic East. Beginning in the 12th century, scientific thought travelled from these centers to Europe. Two streams carried the ideas of Diophantus. One might call them the algebraic and number-theoretic, or arithmetical, respectively. European scholars became familiar with Diophantus' algebraic ideas 300 years before learning his arithmetical ideas. This is not surprising. The new algebra was taken up by Diophantus' Byzantine commentators (M. Planudes and G. Pakhimeres, who lived in the 13th c.) as well as by Arabic mathematicians, especially Abu'l Wafa and his school (10th c.). True, the Arabic mathematicians used words rather than literal symbols for denoting powers of the unknown. Moreover, in working with powers of the unknown they used the awkward multiplicative principle instead of the convenient additive principle used by Diophantus. For example, they called x^6 "square-cube," and not, like Diophantus, "cube-cube." In the case of x^5 they could not devise a name based on the lower powers of the unknown, because 5 is a prime and cannot be written as a product of smaller factors. Hence the name "dumb," or "first inexpressible." The same difficulty arose in connection with all prime powers of the unknown. This notational principle was taken over from the Arabs by European mathematicians. In particular, it was used in Italy during the Renaissance and later by the German algebraists known as cossists. One exception was the gifted 13th-century mathematician Leonardo Pisano, a contemporary of Dante. In his famous "Liber abaci" he not only used the additive principle for powers of the unknown but also was the first European to consider problems which reduced to indeterminate equations.

Diophantus' rules for operating with polynomials and equations were used in the Middle Ages by practically all algebraists.

Negative numbers were adopted far less readily. Arabic mathematicians did not use them at all, and Europeans accepted them with a great deal of scepticism. For a long time they called them "false numbers" and tried to manage without them.

But Diophantus' "Arithmetic" contained a second, far deeper, circle of ideas, namely diophantine analysis. For a long time these ideas were completely unknown. The paradoxical situation which prevailed in Europe in the 15th and 16th centuries was that scholars used and developed the literal algebra derived from Diophantus but knew nothing about his works.

It seems that the first to read Diophantus' works was the 15th-century astronomer Regiomontanus (Johann Müller). While travelling in Italy, Regiomontanus discovered Diophantus' manuscript in Venice and wrote about it to a friend. The content of the manuscript was amazingly rich. Regiomontanus decided to translate it, but first tried to locate all the 13 books mentioned by Diophantus in his introduction. However, only six books, namely the ones known today, were found and the translation was not made.

100 years passed. During that time none of the eminent algebraists, such as Cardano and Tartaglia, knew anything about Diophantus. But in 1572, 143 problems from Diophantus' "Arithmetic" appeared in the "Algebra" of Rafael Bombelli, a professor at the University of Bologna. In the introduction to this work Bombelli writes that "last year a work dealing with this subject was found in the library of our Lord in the Vatican, written by a certain Diophantus, a Greek author who lived at the time of Antoninus Pius." Antoninus Pius was a Roman emperor who lived in the 2nd century AD. It is a mystery what source Bombelli relied on for information about the date of Diophantus' life. After reading the manuscript, Bombelli concluded that its author was "highly knowledgeable in the science of numbers." "In order to enrich the world with a work of such importance," Bombelli and Pacci, the latter a Roman mathematician who found the manuscript, decided to translate it. Bombelli states that "we translated five of the seven books but could not translate the remaining ones because of other obligations." The reference to seven books is baffling. The Vatican manuscript consists of just six books. It is conceivable that the seventh book got lost. If the translation of Bombelli and Pacci had come down to us, then we could compare it with the books in our possession and check whether our assignment of the problems to the respective books agrees with Bombelli's. Unfortunately, the translation has disappeared without a trace.

Bombelli's "Algebra" is remarkable in many respects. It contains improved algebraic notation for powers of the unknown. Complex numbers, $a+bi, i^2=-1$, appeared here for the first time, including precise rules of operation on them. Finally, complex numbers were used in this work to investigate the so-called irreducible case of a cubic equation. For us Bombelli's book is important because it included for the first time Diophantus' problems, admittedly presented out of context.

Diophantus' "Arithmetic" influenced the whole of Bombelli's book. In the original manuscript his own problems were presented in a pseudo-practical form, but in the final version of the book they were formulated abstractly, in the manner of Diophantus. Bombelli also changed some terms and brought them closer to those he found in Diophantus.

The first Latin translation of the "Arithmetic" appeared just three years after the publication of Bombelli's "Algebra." It was prepared by the famous philologist and philosopher Xylander (his true name was Holzmann). On the whole, his translation was a good one, but one senses that its author had no familiarity with mathematics.

In 1585, the problems in the first four books of Diophantus appeared in the book of the well-known mathematician and mechanician Simon Stevin. The second edition of this book, prepared by the gifted algebraist Albert Girard, included the problems in the remaining two books. But Diophantus' methods were fully revived only in the works of the two greatest French mathematicians of the 16th and 17th centuries, François Viète and Pierre Fermat.

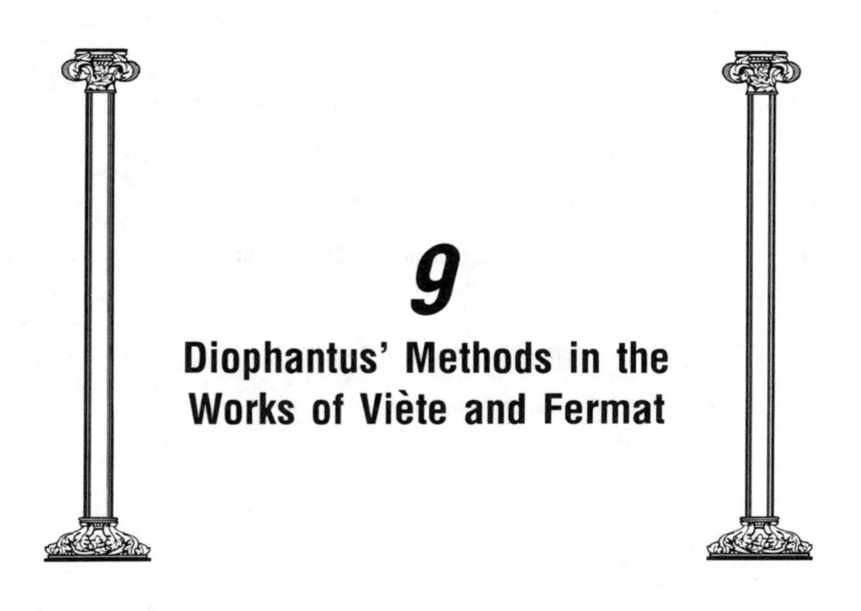

9
Diophantus' Methods in the Works of Viète and Fermat

François Viète is rightly called the father of the literal calculus. Before its appearance one can speak of algebra only in a limited sense. After Diophantus, it was Viète who took a truly new step towards the construction of such a calculus. Viète introduced symbols for the arbitrary constant magnitudes (parameters) in problems. It was only then that the first formulas appeared and it was possible to replace some mental operations with literal ones.

In problem 16 of book V, Diophantus writes: "and we know from the Porisms: The difference of two cubes can be represented as a sum of two cubes." Obviously, what is at issue is the solution of the equation

$$x^3 + y^3 = a^3 - b^3 \qquad (*)$$

for $a > b > 0$ and positive x and y. There is no solution of this problem in the "Arithmetic."

Bombelli proved this porism in his "Algebra." He paid no attention to the method itself. At any rate, he stated no other problems which could be solved by means of this method. In his book with the unusual title "Zetetics" (from the Greek $\zeta\eta\tau\acute{\epsilon}\omega$, to search) Viéte also proved this diophantine porism and posed two additional analogous problems:

$$1. \quad x^3 - y^3 = a^3 + b^3 \qquad (x > y > 0, \quad a > 0, \quad b > 0),$$

$$2. \quad x^3 - y^3 = a^3 - b^3 \qquad (x > y > 0, \quad a > b > 0).$$

He solved all three problems by means of Diophantus' method of tangents. For example, to solve the problem (∗) Viète puts

$$x = t - b, \quad y = a - kt,$$

and obtains after substitution

$$t^3(1 - k^3) + 3t^2(ak^2 - b) + 3t(b^2 - a^2k) = 0.$$

Then he puts $b^2 - a^2k = 0$, which is equivalent to the requirement that the straight line $y = a - k(x + b)$ be tangent to the curve (∗) at $(-b, a)$, and finds that

$$t = \frac{3a^3b}{a^3 + b^3}.$$

For x and y we obtain the expressions

$$x = b \cdot \frac{2a^3 - b^3}{a^3 + b^3},$$

$$y = a \cdot \frac{a^3 - 2b^3}{a^3 + b^3},$$

which show that the solution will be positive only if $a^3 > 2b^3$. This goes against Diophantus' claim, connected with the equation (∗), that "The difference of two cubes can be represented as the sum of two cubes." It was Fermat who managed to solve this riddle, and to overcome a similar difficulty associated with the equation

$$x^3 + y^3 = a^3 + b^3,$$

added by him to the equations of Diophantus and Viéte. If this problem is solved by means of the method of tangents, the difficulty that crops up is that either x or y is negative, that is, the sum of two cubes is represented as the difference, rather than the sum, of two new cubes.

Fermat's idea for overcoming these difficulties was to iterate the method of tangents. In one of his comments on a problem in the "Arithmetic," he considers the equation

$$4x^3 + 6x^2 + 4x + 1 = z^3$$

and says that after the "first operation" one gets $x = -9/22$. Indeed, this is the value obtained by applying Diophantus' method of tangents, that is, by putting $z = \frac{4}{3}x + 1$.

Since x is negative, Fermat applies a "lift" by putting $x = t-(9/22)$, and then applies the method of tangents once more. The lift yields an equation of the form

$$4t^3 + At^2 + Bt + z_1^3 = z^3.$$

If we apply to it the substitution $z = (B/3z_1^2)t + z_1$, then we obtain a positive solution.

This method of Fermat is described in detail in de Billy's "Inventum Novum," where the author stresses that the method yields infinitely many solutions in the case of equations of the form

$$y^2 = f_3(x), \qquad y^3 = f_3(x), \quad \text{and} \quad y^2 = f_4(x),$$

where $f_n(x)$ is a polynomial of degree n with rational coefficients.

To come back to Diophantus' assertion. Can we assume that he had iterated the method of tangents long before Fermat?

In 1621 Bachet de Méziriac published a new translation of Diophantus' "Arithmetic." The new translation was superior to Xylander's and contained the Greek text as well as its Latin translation. This edition became famous not only because of the quality of the translation and Bachet's detailed commentaries but also because Fermat jotted down in his copy of the book his thoughts and results pertaining to number theory. On the margin next to problem 8 of book II, in which Diophantus divides a square into a sum of two squares (see Chapter 5), Fermat wrote: "On the other hand, it is impossible to separate a cube into two cubes, or a biquadrate into two biquadrates, or generally any power except a square into two powers with the same exponent. I have discovered a truly marvellous proof of this, which however the margin is not large enough to contain." (See, for example, [4], p. 108.) This is Fermat's famous Last Theorem which has brought its author fame far beyond the confines of mathematics. This theorem has played an exceptionally important role in the history of mathematics. It was the subject of investigations by Euler, Legendre, Dirichlet, Kummer and other great mathematicians whom it stimulated to construct a new domain of mathematics known as higher arithmetic, or the arithmetic of fields of algebraic numbers.

Who was the author of the Last Theorem? What do we know about him? More than about Diophantus but less than about many of his contemporaries.

Pierre Fermat was born in 1601 in the south of France, near Toulouse, into a prosperous middle-class family. He received a fine education. He had an excellent command of Latin, Italian, and Spanish, and wrote splendid poems in all these languages and in his native French. He knew Greek so well that he corrected many learned translations (including that of the books of Diophantus) and could have acquired fame as an expert on Hellenism. Due to his legal training, he became a councilor of the Parliament (that is tribunal) of the town of Toulouse, where he spent most of his life. Very likely, his life seemed to resemble the lives of his fellow councilors and of his many merchant-relatives. Fermat married, had five children, and seldom left Toulouse. But this orderly and seemingly calm life was tense and stormy. Its core was mathematics, which he came to love by reading the ancient authors, Archimedes, Apollonius, and Diophantus. One of Fermat's first works was his reconstruction of Apollonius' lost treatise "On plane loci," referred to by Pappus. From then on he was possessed by mathematics. This is best seen from his correspondence. Since he lived far from the scientific centers of his time, Fermat was forced to present his results in letters. More than a hundred of these letters have survived. They make compelling reading to this day. They exude their author's passionate desire to know the truths of mathematics, a desire which took hold of him at an early time with irresistible force.

Fermat was undoubtedly the leading mathematician of his time. He created the most general new methods of the part of mathematics which came to be known as infinitesimal analysis. He and Descartes created analytic geometry. He and Pascal created the foundations of probability. Like all scientists of that time, Fermat was very much interested in the application of mathematics to the study of physical phenomena. In particular, he studied optics. Using the minimal principle now named after him he was able to explain how a ray of light moves in an inhomogeneous medium.

Fermat loved number theory. In this area he had no equal. He was able to select from many interesting questions and special problems those fundamental problems whose investigation made number theory a science. Fermat's problems were studied by the greatest mathematicians of the 18th and 19th centuries, beginning with Euler and ending with Hilbert.

We speak of "problems" rather than "theorems" because most of Fermat's assertions have come down to us without proof. They are stated on the margins of his copy of Diophantus' arithmetic or in letters in which he asked fellow scholars to prove them. The one exception is the Last Theorem for biquadrates, whose proof he set down. What Fermat did describe was his new method for proving number-theoretic propositions which he named "the method of infinite or indeterminate descent." We quote from the letter in which he describes his new method:

> ... Since the usual methods presented in books do not suffice to prove such difficult propositions [of number theory], I found a completely new way to achieve this.
>
> I called this method of proof *infinite* or *indeterminate descent*; at first I used it only to prove negative propositions, such as the following:
>
> That there exists no number one less than a multiple of three which is composed of a square and three times a square;
>
> That there exists no right triangle in whole numbers whose area is a square number. The proof is by reductio ad absurdum. If there existed a right triangle in whole numbers whose area were a square, then there would exist another such triangle, smaller than the first, with the same property. Then, by a similar argument, there would exist a third with this property, smaller than the second, and a fourth, fifth, descending to infinity. But given a whole number, there is no infinite descent through smaller numbers. (I have always

in mind whole numbers). Hence the conclusion that there is no right triangle with square area.[1]

We note that the proposition about the area of a triangle whose sides are given by whole numbers, which Fermat uses to demonstrate his method, is equivalent to the proposition that there are no two biquadrates whose difference is a square. A fortiori, there are no two biquadrates whose difference is a biquadrate. But this is the Last Theorem for biquadrates. Its proof by the method of descent, Fermat's only number-theoretic proof, has come down to us. Later Euler used the method of descent to prove the Last Theorem for $n = 3$ and $n = 4$.

Today Fermat's method of descent is an irreplaceable tool in the study of problems of diophantine analysis. But the use of this method in problems which involve rational points on a curve or on some other manifold has necessitated the introduction of the new concept of "height of a point."

For example, let there be given an indeterminate equation

$$f(x, y) = 0, \qquad (*)$$

about which we want to prove that it has no rational solutions. For proof we go over to homogeneous coordinates

$$x = u/z, \quad y = v/z$$

and obtain an equation

$$\Phi(u, v, z) = 0. \qquad (**)$$

To every rational solution of $(*)$ there corresponds an integral solution of $(**)$. Thus it suffices to show that $(**)$ has no integral solutions.

For example, if $(*)$ is of the form

$$Ax^n + By^n = C,$$

then $(**)$ is of the form

$$Au^n + Bv^n = Cz^n.$$

Let (u, v, z) be an integral solution of $(**)$. We define the *height* of the point (u, v, z) as the largest of the numbers $|u|$, $|v|$, $|z|$. To realize the "descent," we must show that if the coordinates of a point of height h satisfy the equation $(**)$, then the coordinates of some other point of height $h_1 < h$ will also satisfy it. Since there are only finitely many positive integers less than h, $(**)$ has no integral solutions, and thus $(*)$ has no rational solutions.

We conclude with a comment on Fermat's treatment of quadratic and cubic indeterminate equations of the form $f(x, y) = 0$. All we can say in this connection is that Fermat understood Diophantus' ideas and skillfully applied his methods, to which he added only the lifting of a curve. Problems which reduce to finding rational solutions of cubic equations are found on the margins of Fermat's copy of the "Arithmetic" as well as in de Billy's work written after Fermat's death with a view to clarifying his methods. In this work, titled "Doctrinae Analyticae inventum novum", which was added to Fermat's collected works (edited by Paul Tannery), Diophantus' methods are applied in a detailed and methodical way, but nothing new has been added to them.

Notes

[1] This letter to Carcavi was published in *Oeuvres de Fermat,* vol. II, p. 43, Paris 1891.

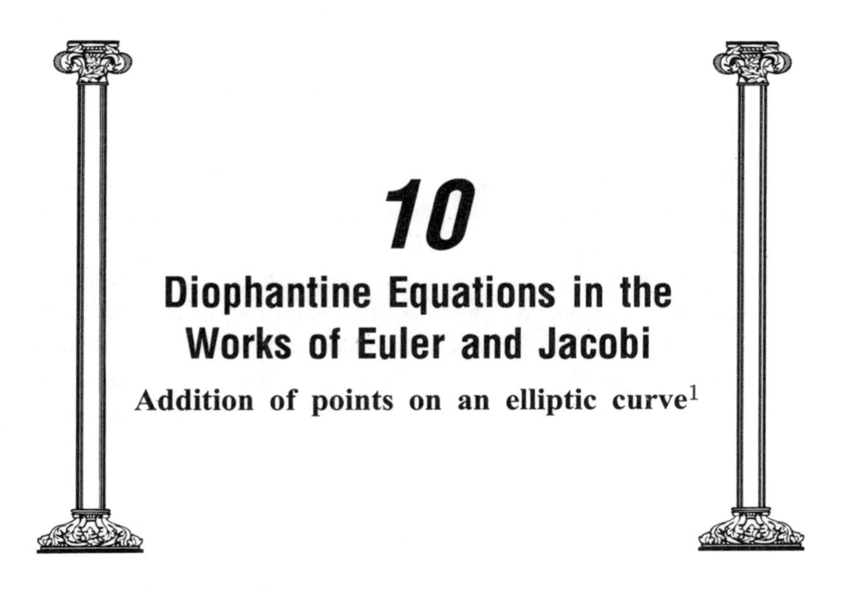

10

Diophantine Equations in the Works of Euler and Jacobi

Addition of points on an elliptic curve[1]

The study of quadratic and cubic indeterminate equations was begun by Diophantus. The first stage of this study was completed by Euler (1707–1783).

Euler, the greatest mathematician of the 18th century, occupies so leading a position in mathematics that there is literally no area of the subject to which he has not contributed fundamental results, deep ideas, or powerful general methods. In particular, this is true of diophantine analysis.

In his "Algebra"[2] Euler systematically analyzed the question of rational solutions of equations of the form

$$y^2 = ax^2 + bx + c \tag{16}$$

and of the form

$$y^2 = ax^3 + bx^2 + cx + d, \tag{17}$$

and gave a precise formulation of the difference between the two cases. Thus he prefaced his investigation of equations of the form (17) with the following observation:

> To begin with, we must state here as well that one cannot, as before, give a general solution. Rather, each operation enables us

to find just one value of x, whereas the method used earlier yields at one time infinitely many solutions.

Moreover, he showed how to obtain a new solution by means of Diophantus' tangent method. His arguments were purely analytical, making no use whatever of geometric terminology.

Euler observed[3] that certain cubic curves behave like quadratic curves, that is, that the unknowns x and y can be expressed as rational functions (with rational coefficients) of a single parameter, and he stated conditions for this to happen. Specifically, in the case of an equation of the form (17), it is necessary that the polynomial on the right have a multiple rational root:

$$\begin{aligned} F_3(x) &= ax^3 + bx^2 + cx + d \\ &= a(x-\alpha)^2(x-\beta). \end{aligned}$$

Euler himself proved the sufficiency of this condition and showed that, in this case, it is possible to obtain rational expressions for x and y by means of the substitution

$$y = k(x-\alpha).$$

Using this substitution we obtain

$$k^2(x-\alpha)^2 = a(x-\alpha)^2(x-\beta),$$

whence

$$\begin{aligned} x &= \frac{k^2 + a\beta}{a}, \\ y &= k\frac{k^2 + a\beta - a\alpha}{a}. \end{aligned}$$

It is easy to show that Euler's condition is equivalent to the curve $y^2 = F_3(x)$ having one double point, that is, its genus is zero. Indeed, the equations

$$y^2 = ax^3 + bx^2 + cx + d,$$

$$3ax^2 + 2bx + c = 0,$$

$$2y = 0,$$

which define the singular points, imply that the abscissa of a double point must be a root of the polynomial

$$F_3(x) = ax^3 + bx^2 + cx + d$$

and of its derivative

$$F_3'(x) = 3ax^2 + 2bx + c = 0,$$

and thus a multiple root of $F_3(x)$. The fact that this root can be found by applying the Euclidean algorithm to $F_3(x)$ and $F_3'(x)$ shows that it must be rational.

Later Poincaré showed that Euler's condition is not only necessary but also sufficient (see Chapter 12).

In the last years of his life Euler returned to diophantine analysis. He perfected his methods and for the first time applied Diophantus' secant method for two given rational points on the curve (17). Specifically, let

$$F_3(\alpha) = f^2, \quad F_3(\beta) = g^2. \tag{18}$$

Euler put

$$y = f + \frac{g-f}{\beta-\alpha}(x-\alpha)$$

or

$$y = g + \frac{f-g}{\alpha-\beta}(x-\beta),$$

which is equivalent to passing a straight line through the points (α, f) and (β, g), and obtained a new rational value of x from the equation

$$F_3(x) = \left[f + \frac{g-f}{\beta-\alpha}(x-\alpha)\right]^2.$$

For this one needs only bear in mind the equalities (18).

These papers were published only in 1830, that is, after Euler's death.

Euler was the author of certain other investigations, at first glance unrelated to the problems of Diophantus, that contributed a completely new viewpoint to the treatment of these problems. We have in mind the famous theorem, discovered by Euler, on the addition of elliptic integrals.

Let

$$y^2 = ax^3 + bx^2 + cx + d \tag{19}$$

be a given curve[4]—call it Γ—and let $A(x, y)$ be a point on Γ. Put

$$\Pi(A) = \int_{\infty}^{x} \frac{dx}{y}.$$

Euler's first theorem is the assertion that for arbitrary points $A(x, y)$ and $B(x_1, y_1)$ on Γ there is a point $C(x_2, y_2)$ on Γ such that

$$\Pi(A) + \Pi(B) = \Pi(C), \tag{20}$$

and such that the coordinates of C are rationally expressible in terms of the coordinates of A and B (that is, as rational functions with rational coefficients[5]).

Euler's second theorem is the assertion that if A and D are points on Γ and n is a nonzero integer such that

$$\Pi(D) = n\Pi(A), \tag{21}$$

then the coordinates of D are rationally expressible in terms of the coordinates of A. In particular, for $n = 2$ we have

$$\Pi(D) = 2\Pi(A). \tag{22}$$

The relation (21) is sometimes referred to as the *theorem on the multiplication of elliptic integrals.*

If A and B are rational points, then so are C and D. This means that Euler's theorem enables us to obtain from one or two rational points on Γ new rational points on Γ.

The first to note the connection between Euler's addition theorem and diophantine analysis was the famous German mathematician Carl Gustav Jacob Jacobi. He did this in the paper "On the use of elliptic and abelian integrals in diophantine analysis" (De usu theoriae integralium ellipticorum et integralium abelianorum in analysi Diophantea), published in *Crelle's Journal,* the most important 19th-century German mathematical journal, in 1834. It seems that Jacobi's contemporaries ignored the paper in spite of its profound and interesting content.

At the beginning of his paper Jacobi expresses surprise that the "learned man" (that is, Euler) overlooked the connection which he, Jacobi, is about to discuss and regards as obvious. He then formulates the Euler addition theorem (for the case of two as well as one given point) and notes that given a finite number of rational points $A_1, \ldots, A_s$ on the curve Γ one can obtain infinitely many new rational points on Γ from the relation

$$\Pi(A) = m_1\Pi(A_1) + \cdots + m_s\Pi(A_s),$$

where $m_1, \ldots, m_s$ are arbitrary integers. Similarly, starting with a single rational point and using the relation (22) we can obtain an infinite sequence of rational points. However, if we assign to n the values $\pm 2, \pm 3, \ldots$ we need not always obtain new points. It can happen that for some n

$$n\Pi(A) = \Pi(A),$$

that is, after a finite number of steps we may return to the initial point. Jacobi is aware of this possibility and finds the determining condition (but to state this condition we would have to immerse ourselves in the study of the periods of the integrals $\Pi(A)$, contrary to our intentions). We will call points for which there is an n such that

$$n\Pi(A) = \Pi(A)$$

points of finite order.

At the end of his paper Jacobi indicates how to extend these results to algebraic curves of higher order by replacing Euler's addition theorem with the more general theorems of Abel. Without going into these questions we note that these ideas of Jacobi have been developed only in our own time.

We go back to the essential content of the paper and show that here Jacobi came close to discovering the structure of the set of rational points on an elliptic curve. To make this discovery he lacked not the technical means—of which he had a masterly command—but a new viewpoint, one that made its way only gradually and with difficulty in the last century. We will try to clarify the essence of this viewpoint. To do this we consider the set $\mathcal{M}$ of rational points on the curve Γ. If A

and B are two points in $\mathcal{M}$ then, by Euler's theorem, there is a point C in $\mathcal{M}$ such that

$$\Pi(A) + \Pi(B) = \Pi(C).$$

We will view C as the "sum" of A and B and write

$$A \oplus B = C.$$

We write $\oplus$ rather than + to emphasize that we are not dealing with the addition of numbers. Thus we have defined on $\mathcal{M}$ a binary operation, that is, a rule of composition that associates with any two elements A and B in $\mathcal{M}$ a third element C in $\mathcal{M}$.

In modern mathematics we call a set $\mathcal{S}$ with a law of composition $\oplus$ a group provided that the following conditions hold:

1. For any three elements A, B, and C in $\mathcal{S}$

$$(A \oplus B) \oplus C = A \oplus (B \oplus C) \quad (\textit{associativity}).$$

2. $\mathcal{S}$ contains an *identity element* N such that for all A in $\mathcal{S}$

$$A \oplus N = A.$$

3. For every element A there is an *inverse* element A' in $\mathcal{S}$ such that

$$A \oplus A' = N.$$

If, in addition, for any two elements A and B in $\mathcal{S}$

$$A \oplus B = B \oplus A,$$

then the group is called *commutative* or *abelian*. Thus the integers form an abelian group under addition, the positive rational numbers form an abelian group under multiplication, and all two-by-two real matrices with nonzero determinant form a noncommutative group under multiplication; in the latter example the unit matrix plays the role of the identity element.

By analogy with operations on numbers we usually refer to the group operations as *addition* or *muliplication*. This being so, it is natural to speak of additive inverses and multiplicative inverses. The term "addition" is usually used in connection with commutative groups. In such groups the identity element is sometimes called the zero element.

We wish to find out if the set of points $\mathcal{M}$ with the operation $\oplus$ is a group.

The associativity of our operation follows from the associativity of the addition of integrals, where, obviously,

$$\left[\Pi(A) + \Pi(B)\right] + \Pi(C) = \Pi(A) + \left[\Pi(B) + \Pi(C)\right].$$

It remains to see if $\mathcal{M}$ contains an element that plays the role of zero and if for each element in $\mathcal{M}$ there is an additive inverse in $\mathcal{M}$.

We begin with the zero element. If $\mathcal{M}$ is the set of rational points on Γ, that is, points whose coordinates are finite rational numbers, then $\mathcal{M}$ has no zero point. To speak of addition of points in $\mathcal{M}$ we must supplement it with an extra point $\mathcal{O}$ that will play the role of zero. In the next chapter we explain how this is done and for the time being take the matter on faith. Clearly, for such an $\mathcal{O}$ we must have

$$\Pi(\mathcal{O}) = 0.$$

Now we can find for every A its additive inverse. It is natural to say that $A \oplus A' = \mathcal{O}$ if

$$\Pi(A) + \Pi(A') = 0.$$

But then for A' we must take the point that is symmetric to A with respect to the x-axis. Indeed, if A has coordinates (x, y), then A' has coordinates $(x, -y)$ and

$$\begin{aligned} \Pi(A') &= \int_{\infty}^{x} \frac{dx}{-y} \\ &= -\int_{\infty}^{x} \frac{dx}{y} \\ &= -\Pi(A). \end{aligned}$$

We note that

$$\Pi(A) + \Pi(B) = \Pi(B) + \Pi(A),$$

that is

$$A \oplus B = B \oplus A.$$

This means that our group is abelian.

Thus beginning with Euler's theorem and using the connection noted by Jacobi it is possible to make the set of rational points on an elliptic curve, with one additional point, into a commutative group. The points of finite order mentioned by Jacobi become the group elements of finite order (a group element is said to be of finite order n if n times that element is the zero element of the group). This is a modern translation of Jacobi's observations.

The mathematicians of the first half of the 19th century did not think of extending arithmetical operations to points or other objects very different from numbers. This being so, Jacobi "added" not points A and B but integrals $\Pi(A)$ and $\Pi(B)$, with addition of integrals given its usual interpretation.

Notes

[1] An algebraic curve of genus 1 is called *elliptic.*

[2] This book was first published in Russian under the title "Universal arithmetic" (1768–1769). Later there appeared many German and French translations of it.

[3] This was already noted by Diophantus. Thus problem 6 in book IV can be reduced to the equation

$$x^3 + 16x^2 = y^3,$$

which yields x and y as rational functions of a single parameter:

$$x = 16/(a^3 - 1),$$
$$y = ax = 16a/(a^3 - 1).$$

[4] Euler considered the theorem for curves $y^2 = F_3(x)$ and $y^2 = F_4(x)$. We limit ourselves to the first case. The second case can be reduced to the first.

[5] Euler found explicit expressions for these functions. Thus, given the coordinates of A and B, one can compute the coordinates of C.

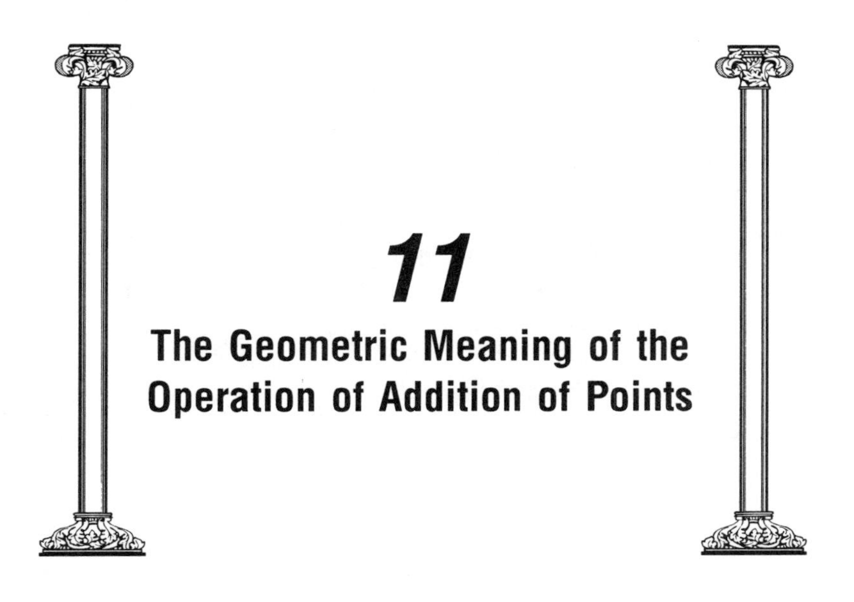

11 The Geometric Meaning of the Operation of Addition of Points

We pose the following question: is there a connection between Euler's addition theorem and Diophantus' tangent and secant methods? After all, in both cases we start with one or two rational points on the curve Γ and obtain new rational points on Γ. Neither Euler nor Jacobi refer to such a connection—and yet one exists!

Let us make our question more precise. By Euler's theorem, given two points A and B on Γ we can find a point C on Γ such that

$$\Pi(C) = \Pi(A) + \Pi(B).$$

On the other hand, let us pass a line through A and B and find its intersection C' with Γ. Is there a connection between C and C'? The answer is yes and the connection is very simple. The two points are symmetric with respect to the x-axis, that is, if C has coordinates (x_2, y_2) then C' has coordinates $(x_2, -y_2)$.

Now the geometric sense of the operation of addition of points on a curve is obvious: the sum of points A and B on Γ is the point C on Γ symmetric, with respect to the x-axis, to the point of intersection of Γ with the line AB.

The limitation of this method is that we cannot use it to add A to itself, that is, to obtain $2A$. To get around this difficulty we proceed thus.

Following Diophantus' first method, we pass through A the tangent to Γ at A and find D', its point of intersection with Γ. The point D' is symmetric to the point D obtained by Euler's method: $\Pi(D) = 2\Pi(A)$. This means that we can determine the point $2A$, and, more generally, the point nA for all n by purely geometric means.

With this interpretation of addition, which point will play the role of zero? To answer this question we proceed as in Chapter 6, that is, we go over to homogeneous coordinates. If we put

$$x = u/z, \quad y = v/z,$$

then equation (19) becomes

$$v^2 z = u^3 + auz^2 + bz^3. \tag{19'}$$

Equation (19′) shows that if $z = 0$ then $u = 0$ and v is arbitrary. Since the (homogeneous) coordinates are determined up to a multiplicative constant, we take $v = 1$.

We assume that the point at infinity on our curve corresponds to the triple $(0, 1, 0)$ and denote that point by $\mathcal{O}$. We further assume that $\mathcal{O}'$, the point symmetric to $\mathcal{O}$ with respect to the x-axis, coincides with $\mathcal{O}$.

We will show that the point $\mathcal{O}$ plays the role of zero. We note that all vertical lines $u = cz$ intersect at $\mathcal{O}$. Indeed, $z = 0$ implies $u = 0$, and v can be taken to be equal to 1.

Now let A be a rational point on the curve Γ with coordinates (x_0, y_0). By what has just been proved, the line through A and $\mathcal{O}$ is vertical, that is, its equation is

$$x = x_0.$$

This line will intersect the curve Γ in three points, namely A, $\mathcal{O}$, and the point $A'(x_0, -y_0)$, symmetric to A with respect to the x-axis. According to our definition, the sum of A and $\mathcal{O}$ is the point symmetric to A', that is, A itself. This means that

$$A \oplus \mathcal{O} = \mathcal{A}.$$

Finally, the additive inverse of A is $A'(x_0, -y_0)$. To see that this is so note that the line joining A and A' is vertical, and thus intersects Γ

at $\mathcal{O}$. By definition, the sum of A and A' is the point symmetric to $\mathcal{O}$, which, by assumption, coincides with $\mathcal{O}$. Thus

$$A \oplus A' = \mathcal{O}.$$

We note that our point $\mathcal{O}$ has the property

$$\Pi(\mathcal{O}) = \int_{\infty}^{\infty} \frac{dx}{y} = 0.$$

Thus Euler's addition theorem also implies that the role of zero must be played by a point at infinity.

We see that the "addition" of points on an elliptic curve can be based on Diophantus' procedures. Were Euler and Jacobi aware of this? More precisely, did they know that three points on Γ with

$$\Pi(A) \oplus \Pi(B) \oplus \Pi(C) = 0$$

are collinear? Neither Euler nor Jacobi mention this but Euler may have known it, and Jacobi should have known it very well indeed. On the other hand, both formulated the addition theorem for a curve

$$y^2 = ax^4 + bx^3 + cx^2 + dx + e \qquad (*)$$

without bothering with cubic curves. And for curves $(*)$ with rational points the addition theorem has no simple and uniquely determined geometric sense. Furthermore, neither Euler nor Jacobi attached significance to the geometric interpretation of analytic expressions.

In spite of the simplicity of the reasoning underlying the "addition" of points on an elliptic curve, it was about 70 years before this reasoning became the basis for a systematic study of the structure of the set of its rational points. This was done at the beginning of this century by the great French mathematician Henri Poincaré.

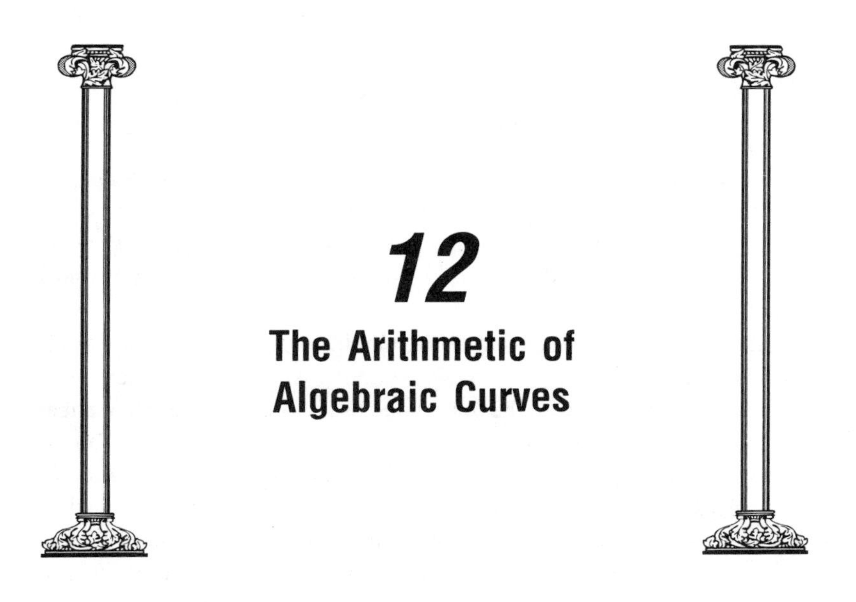

12
The Arithmetic of Algebraic Curves

Jacobi's paper remained unnoticed, and the first to conceive of the idea of constructing an arithmetic on an elliptic curve was Poincaré. Nevertheless, a great deal of work was done between 1834 and the end of the 19th century in the study of the geometry of algebraic curves. The concept of genus of an algebraic curve (see Chapter 3) first appeared in the works of the eminent Norwegian mathematician Niels Henrik Abel (1802–1829). Beginning with different considerations, Bernhard Riemann (1826–1866), Germany's greatest mathematician, was led to the same concept. In his remarkable paper "The theory of abelian functions" (1857) Riemann based the classification of equations of the form $F(s, z) = 0$ on birational transformations—which he referred to as *rational substitutions*—and showed that *the genus of a curve is invariant under such transformations.* Riemann wrote:

> We will regard all irreducible algebraic equations in two unknowns that can be transformed into one another by means of rational substitutions as belonging to the same class; thus the equations $F(s, z) = 0$ and $F_1(s_1, z_1) = 0$ belong to the same class if s and z are rationally expressible in terms of s_1 and z_1 in such a way that $F(s, z) = 0$ goes over into

$F_1(s_1, z_1) = 0$, and s_1 and z_1 are likewise rationally expressible in terms of s and z. [6]

In successive papers Clebsch and other German mathematicians laid the foundations of a theory of algebraic curves. As a rule, these curves were considered over the complex numbers (that is, their coefficients were complex numbers) and this ruled out consideration of their arithmetic.

Henri Poincaré begins his memoir "On the arithmetic properties of algebraic curves" (*Journ. de mathém. pures et appl.,* Paris, 5-me série, v. 7 (1901) pp. 161–234) with the important observation that the arithmetical properties of many objects are very closely related to their transformations: thus, for example, in the case of quadratic forms in two variables the relevant transformations are—as Gauss showed—linear substitutions with integer coefficients. "We can assume (writes Poincaré) that the study of analogous groups of transformations will be of great help to Arithmetic. This has induced me to publish the following observations even though they are a program of study rather than a genuine theory."

Poincaré began to think of ways of connecting and systematizing the problems of diophantine analysis. To this end he decided to carry out a new classification of polynomials in two variables with rational integral coefficients. He based this classification on the totality of birational transformations with rational coefficients. We mentioned earlier that Riemann introduced an analogous classification. The difference is that Riemann considered birational transformations with complex coefficients rather than—as did Poincaré—birational transformations with rational coefficients. This fact brought Poincaré to the study of arithmetic properties of curves. We pass to some specific technical aspects of Poincaré's work. According to him, two curves

$$f_1(x, y) = 0 \quad \text{and} \quad f_2(x, y) = 0$$

are equivalent or *belong to the same class* if it is possible to go from one to the other by means of a birational transformation with rational coefficients. (Note that this definition is the same as the definition of birational equivalence given in Chapter 3.) For example, two lines

$$ax + by + c = 0$$

and

$$a'x + b'y + c' = 0$$

with rational coefficients are equivalent.

To prove this Poincaré chooses a rational point F off both lines and associates with each point A on the first line the point A' at which the line AF meets the second line (Figure 4). This implies that all lines with rational coefficients form a single class.

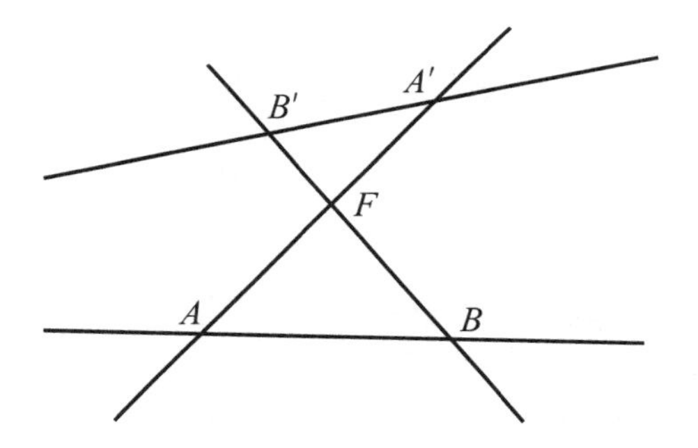

FIGURE 4

Next Poincaré investigates conics, that is, quadratic curves, and shows that if a conic $f(x, y) = 0$ (with integral or rational coefficients) has at least one rational point, then it is equivalent to a rational line. To this end he associates with each point A of a fixed rational line L the point A' on the conic Γ such that A, A', and C are collinear (see Figure 5). We saw that this result had already been established by Diophantus.

Next Poincaré considers cubic curves of genus 0. Since, by the definition of genus,

$$0 = \frac{(3-1)(2-1)}{2} - d,$$

it follows that $d = 1$, that is, the curve has one double point. Poincaré claims that this point must be rational. We will not prove this fact here. But we do want to remind the reader that Euler established the fact

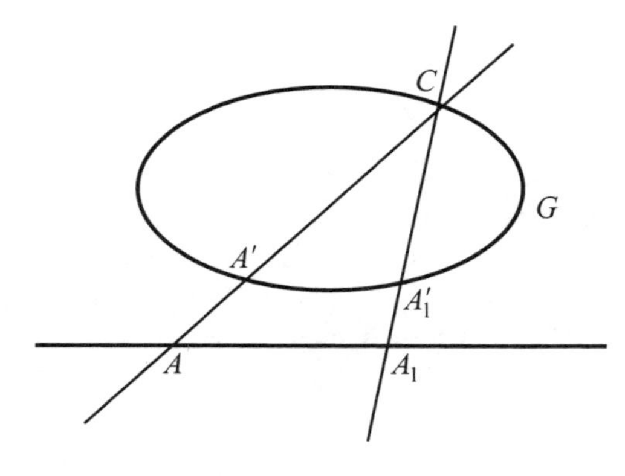

FIGURE 5

that the existence of a double point on a curve $y^2 = F_3(x)$ implies the possibility of expressing x and y as rational functions of a single parameter (see Chapter 10).

Euler did not prove the necessity of this condition. Poincaré's proof that the existence of a double point on a cubic curve is sufficient for the cubic to be equivalent to a line is the same as Euler's, except that Poincaré's is geometric rather than analytic. He takes the double point as a basic point, fixes a rational line L, and associates with each point A on L the point A' on Γ that lies on the line AC (see Figure 6).

Next Poincaré proves a basic theorem that completely characterizes the set $\mathcal{M}$ of rational points on a curve of genus 0.

Theorem. *Every curve of genus* 0 *and order* m, $m > 2$, *is birationally equivalent to a curve of order* $m - 2$.

Thus every curve of genus 0 and odd order $(m = 2k + 1)$ is birationally equivalent to a line, and every curve of genus 0 and even order $(m = 2k)$ is birationally equivalent to a conic.

In particular, this means that a curve of genus 0 and odd order has infinitely many rational points.

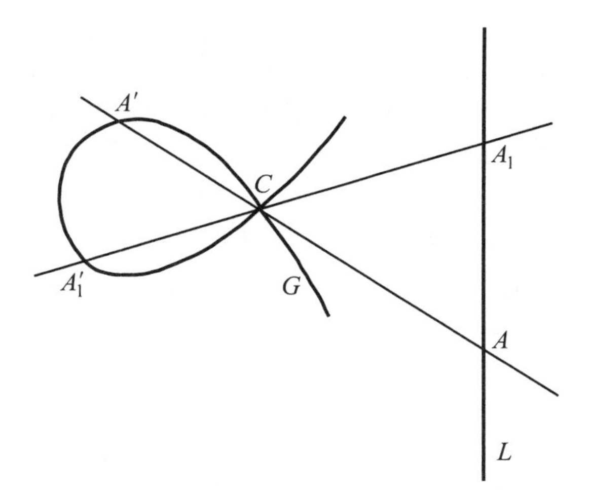

FIGURE 6

The question of rational points on curves of genus 0 and even order reduces to the determination of the rational points on a conic. The structure of this set had already been investigated by Diophantus.

We note that similar results pertaining to curves of genus 0 were obtained ten years earlier by D. Hilbert and A. Hurwitz in the paper "On diophantine equations of genus 0" (*Acta Math.* 14 (1890)). Its authors noted the invariance of the set of rational points of an algebraic curve under birational transformations with rational coefficients. Poincaré seems not to have known this paper; at any rate he does not mention it. In any case, the core of his memoir is devoted not to these facts but to the investigation of curves of genus 1.

He begins by considering the simplest curves of genus 1, that is, cubics. If such a curve has at least one rational point, then, as noted earlier, its equation can be reduced by birational transformations to one of the form

$$y^2 = x^3 + ax + b. \qquad (*)$$

We assume that this has already been done. Poincaré describes Diophantus' tangent and secant methods (without, of course, mentioning his name) for finding new rational points on Γ given, respectively, one and two rational points on Γ. He first formulates both methods geometrically and then connects them with Euler's addition theorem by noting (as Jacobi failed to) that three points A, B, and C on an elliptic curve Γ such that

$$\Pi(A) + \Pi(B) + \Pi(C) = 0 \qquad (**)$$

are collinear. Poincaré also makes more precise the meaning of this equality; we explain this next.

The integral

$$\int_a^u \frac{dx}{y},$$

where y is determined by $(*)$, is an infinitely-many-valued function of its upper limit u. In this it resembles the function

$$\int_o^u \frac{dx}{\sqrt{1-x^2}} = \sin^{-1} u,$$

whose "principal values" lie in the interval $(-\pi/2, \pi/2)$ and whose other values are obtained by adding to the principal values multiples of 2π.

Similarly, the integral

$$\int_o^u \frac{dx}{\sqrt{x^3+ax+b}}$$

has "principal values" and all remaining values differ from the principal ones by terms of the form $m_1\omega_1 + m_2\omega_2$, where m_1 and m_2 are integers and ω_1 and ω_2 are periods whose ratio is a complex nonreal number. Since there are three integrals in $(**)$ and each is determined up to a period, the summands can be chosen so that their sum is 0. This element of precision introduced by Poincaré was neglected by Euler and Jacobi, who tended to take a formal view of mathematical relations. Such precision became a standard characteristic of works dating from the beginning of the 20th century.

Poincaré gives an explicit definition of addition of rational points on an elliptic curve Γ and shows that the set $\mathcal{M}$ of these points forms a commutative group. The definitions of addition and of doubling of points alone make it clear that if $A_1, \ldots, A_s$ are in $\mathcal{M}$ then so is

$$A = m_1 A_1 + \cdots + m_s A_s. \qquad (***)$$

Poincaré asks: is it possible to choose points $A_1, \ldots, A_s$ so that all rational points on Γ are obtainable from the expression $(***)$? In the language of group theory: is the group of rational points on Γ finitely generated? Thus Poincaré initiated a deeper study of the structure of the set $\mathcal{M}$.

Poincaré calls points $A_1, \ldots, A_s$, from which one can obtain all the others by rational operations, a *fundamental system of rational points.* He notes that a fundamental system can be chosen in infinitely many ways, and chooses one with the least number of points. He calls this minimum number (r, say) the *rank* of the curve Γ.[1] It can be shown that the rank is invariant under birational transformations and is thus one of the fundamental intrinsic properties of the curve.

Poincaré asks: "What are the possible values of the integer that we have called the rank of a cubic curve?

This question was interpreted by later mathematicians as the assertion of the finiteness of the rank of an elliptic curve, that is, the assertion that the group of rational points of an elliptic curve has a finite number of generators. This assertion came to be known as the Poincaré conjecture. It was proved only in 1922 by the English mathematician L.J. Mordell. This was the most impressive result since Poincaré's own. In proving that the rank of a curve of genus 1 is always finite, Mordell used Fermat's method of infinite descent.

After discussing cubics Poincaré discusses other curves of genus 1. He proves the following result:

Let $f(x, y) = 0$ *be a curve of genus 1 and order* m. *If there is at least one rational point on this curve, then it is birationally equivalent to a cubic.*

This result settles completely the question of curves of genus 1: such a curve either has no rational points or is equivalent to a cubic. In the latter case its set of rational points has the same structure as the set of rational points of the curve in $(*)$.

Poincaré's memoir contains other interesting ideas and "study programs" but we cannot go into these here. We do wish, however to note a fact of interest for the history of mathematics: it seems that Poincaré was completely unaware of the work of his predecessors bearing on the arithmetic of algebraic curves. He knew of Diophantus' procedures and of their connection with Euler's addition theorem from the general theory of algebraic curves. (Rational points are not the only things that can be added! After all, rational points possess no special geometric distinction.) But the idea of using known facts and methods to study arithmetic properties of curves was conceived by Poincaré independently of others. Thus this idea arose at least three times: in the middle of the third century A.D. in the work of Diophantus, in the 1830s in the work of Jacobi, and at the beginning of the 20th century in the work of Henri Poincaré. This is not a unique occurrence in the history of mathematics. Thus projective geometry was discovered three times: once in antiquity, a second time in the 17th century by Desargues and Pascal, and, "for the last time," at the beginning of the 19th century by Poncelet and others. In saying "for the last time" we mean that from then on there has been no break in the continuity and tradition of these investigations. The same is true of the arithmetic of algebraic curves after Poincaré.

Notes

[1] "Theorie der Abel'schen Functionen" (Borchard's *Journal fuer reine und angewandte Mathematik,* Bd.54, 1857). In: *Riemann's Collected Works,* B.G. Teubner, Leipzig, 1892, p. 119.

[2] Today one defines the rank of an elliptic curve as the least number of rational points $A_1, \dots, A_r$ such that every rational point A on the curve can be written as

$$A = m_1 A_1 + \cdots + m_r A_r + P,$$

where P is a point of finite order.

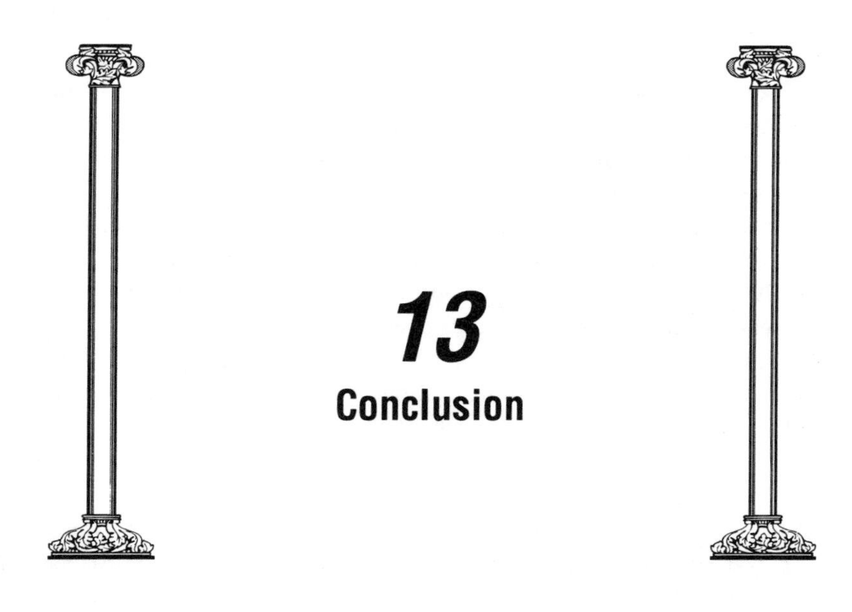

13 Conclusion

We shall now describe some of the generalizations, results, and conjectures pertaining to the arithmetic of algebraic curves. One such generalization is already found in Poincaré's memoir. From Diophantus to Poincaré the arithmetical properties of curves have been considered over the field of rationals, that is, the coefficients of the equation

$$f(x, y) = 0$$

of the curve Γ, the coefficients of all birational transformations, and the coordinates of points to be found must all belong to the field Q of rational numbers. Poincaré suggested that similar investigations be carried out over fields of algebraic numbers, for example over quadratic fields $Q(\sqrt{D})$. In that case a point is called rational if its coordinates belong to the field in question.

Of course, it is possible to construct an arithmetic of curves over an arbitrary field k—say, over the field of rational functions of one variable or over a finite field (of residues modulo p).

In 1929 the French mathematician André Weil used Fermat's method of infinite descent to prove the Poincaré conjecture of the finiteness of the rank of an elliptic curve over an arbitrary field k.

Another generalization, also initiated by Poincaré, pertains to the arithmetic of curves of genus $p > 1$. In this case it is no longer possible to define addition of points, but it is possible to define "addition" for sets of p points, where p is the genus of the curve. Such an addition is already implicit in the previously discussed paper of Jacobi and was discussed by Poincaré in the last paragraph of his memoir. In his 1929 paper Weil proved the conjecture of finiteness of the rank of an algebraic curve of arbitrary genus over an arbitrary field k.

A parallel development involved the question of integral points (that is, points with integral coordinates) on an algebraic curve. Already in 1923 Mordell showed that the equation

$$Ey^2 = Ax^3 + Bx^2 + Cx + D$$

has only a finite number of integral rational solutions. Here the most general result was obtained by the German mathematician C.L. Siegel, who used the methods of A. Thue and those of Mordell and Weil to prove the finiteness of the number of integral points on a curve of genus $p > 0$ over the field k of algebraic numbers.

Translator's note. The material that follows brings this chapter up to date. It was written by Prof. J.H. Silverman of Brown University.

As for rational points on a curve of genus $p > 1$, Mordell conjectured in 1922 that the number of such points is finite. This conjecture was settled in 1983 by Gerd Faltings, who showed that any algebraic curve of genus $p > 1$ defined over a number field K has a finite number of K-rational points. Subsequently Paul Vojta re-proved Faltings' theorem by an entirely different method, and Faltings then extended Vojta's ideas to prove finiteness of rational and integer points on certain subvarieties of abelian varieties, thereby vastly generalizing Siegel's and his own earlier results.

We note that all theorems bearing on systems of generators for the group of rational points on an elliptic curve are existence theorems. We know of no effective method for finding generators, although Yuri Manin has pointed out that an effective method may be deduced from certain widely believed conjectures. The question (posed by Poincaré) of the values that can be taken on by what Poincaré calls the rank of an elliptic curve remains open. Various mathematicians, including especially André Néron and J.-F. Mestre, have given methods for finding elliptic curves with moderately large rank; the current record (1996) is a curve of rank 21 discovered by K. Nagao and T. Kouya. In the positive direction, A.I. Lapin (in characteristic 0) and I.R. Shafarevich (in characteristic p) have shown that there are elliptic curves over the field of rational functions whose ranks are arbitrarily large.

In the past 30 years much of the deepest work on the number-theoretic properties of elliptic functions has been fueled by two conjectures. The first, formulated by B. Birch and P. Swinnerton-Dyer, suggests that the existence of rational points on elliptic curves is closely related to the values of a certain complex analytic function, called an L-series, which encodes information about the points modulo p on the elliptic curve for all primes p. Major progress on this conjecture has been made by J. Coates, B. Gross, V.A. Kolyvagin, K. Rubin, A. Wiles, D. Zagier, and others, but much remains to be done. The second conjecture, formulated by Y. Taniyama and refined by G. Shimura and A. Weil, says that every elliptic curve can be parametrized by certain complex analytic functions, called modular functions, which satisfy various complicated transformation laws. This Modularity Conjecture has now largely been proved by Andrew Wiles, with assistance from Richard Taylor and an extension by Fred Diamond.

Wiles' work on the Modularity Conjecture was motivated, at least in part, by an observation of Gerhard Frey relating the Modularity Conjecture to Fermat's Last Theorem (Chapter 9). Frey suggested that if

$$u^n + v^n = z^n$$

has a solution in nonzero integers with $n \geq 3$, then the elliptic curve

$$E: \ y^2 = x(x - u^n)(x + v^n)$$

would not satisfy the Modularity Conjecture. After Jean-Pierre Serre gave a more precise formulation of Frey's idea, Ken Ribet proved that such an E could not satisfy the Modularity Conjecture. Thus a spectacular corollary of Wiles' theorem was a proof, approximately 350 years after having been penned by Fermat in the margin of his copy of Diophantus' "Arithmetic," that his Last Theorem is indeed true.

It is proper to mention in this short account the contributions of two Russian world-class algebraic geometers. They are Alexei Parshin and Sergei Arakelov. Parshin has many important papers. One of his most important gave a new proof of the Mordell conjecture over function fields, and it was by following an analogous path that Faltings gave the first proof over number fields. Arakelov is best known for "Arakelov intersection theory," in which he described a local intersection pairing at the infinite (archimedean) place of a number field, thereby establishing a close and fruitful connection between the number theory of curves and the algebraic geometry of projective surfaces.

However, we cannot prove—nor even state—the conjectures and results mentioned in this chapter without extensive use of modern algebra and algebraic geometry. To do this would not only take us beyond the technical scope of this small book, but would also force us to study current problems in diophantine analysis rather than its history. Readers interested in the former could consult some of the references given in the bibliography.

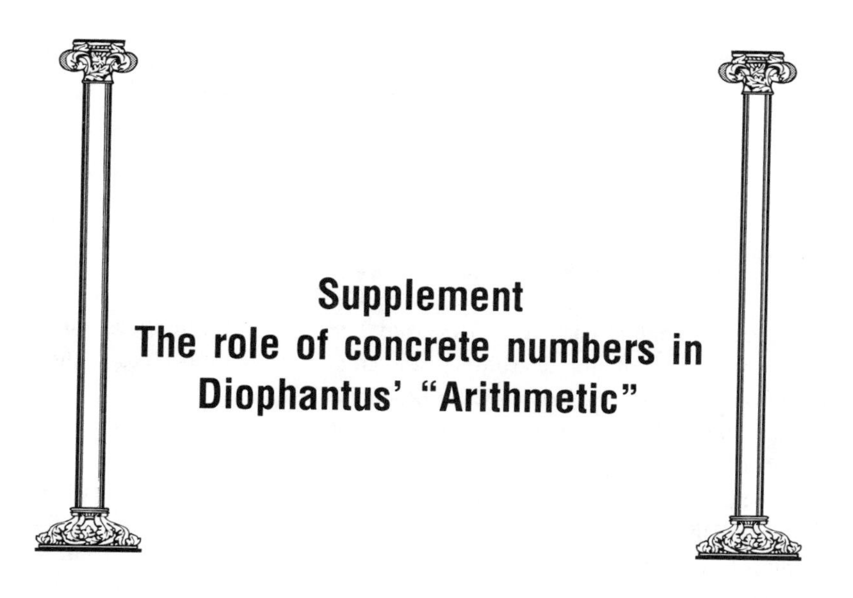

Supplement
The role of concrete numbers in Diophantus' "Arithmetic"

Already in the mathematics of ancient Babylon, the solution of an algebraic problem with concrete numerical data served two different purposes. One was to obtain a numerical solution of the problem. The other was to illustrate an algorithm for the solution of a class of problems of the same type.

In Diophantus' "Arithmetic" the role of numerical parameters was increased in a fundamental way. As a rule, in order to solve a problem Diophantus represented the required numbers as rational functions of a single unknown and of parameters. He assigned to the parameters concrete numerical values but stipulated that these could be replaced by other arbitrary rational numbers, or by arbitrary numbers satisfying certain conditions. As an illustration, we consider problem 8 in book 2, which deals with the representation of a given square as a sum of two squares, i.e., with the solution of the equation

$$X^2 + Y^2 = a^2.$$

Diophantus puts $a^2 = 16$. He takes the base of one of the squares as the unknown $X = t$ and the base of the other square as a linear function of t: $Y = kt - 4$. Here 4 is a root of 16 and k can be an arbitrary rational number. Diophantus writes that it is necessary to take "a certain quantity $t \ldots$ let it be 2." The solution of the problem is given by

$$X = t = \frac{2ak}{1+k^2},$$

$$Y = kt - a = a\frac{k^2-1}{k^2+1}.$$

In Diophantus' case $X = 16/5$ and $Y = 12/5$. But he understands perfectly well that for an arbitrary rational k one obtains a corresponding rational solution. For example, in problem 19 in book III he writes that "we know that every square can be decomposed into two squares in infinitely many ways."

Thus in problem 8 in book II the number 2 performs two functions, that of the concrete number 2 and that of a symbol which stands for an arbitrary rational number. But it is not always possible to assign to a parameter an arbitrary value. For example, in problem 8 in book IV, which is equivalent to the system of equations

$$X_1^3 + X_2 = Y^3,$$

$$X_1 + X_2 = Y.$$

Diophantus puts initially $X_2 = t$, $X_1 = kt$, where $k = 2$. Then $Y = (k+1)t$ and

$$t^2 = \frac{1}{(k+1)^3 - k^3}.$$

For $k = 2$ this yields $t^2 = 1/19$, i.e., t is not rational. To obtain a rational solution one must find two numbers that differ by 1 such that the difference of their cubes is a square:

$$(\tau + 1)^3 - \tau^3 = \square,$$

or

$$3\tau^2 + 3\tau + 1 = \square.$$

Diophantus puts

$$\square = (1 - \lambda\tau)^2$$

and obtains

$$\tau = \frac{3 + 2\lambda}{\lambda^2 - 3}.$$

By choosing $\lambda = 2$ he obtains $\tau = 7$. Hence the value of the parameter can be chosen from the class of numbers

$$\left\{ \frac{3 + 2\lambda}{\lambda^2 - 3} \right\}.$$

Thus we see that in Diophantus' algebraic formalism, in addition to symbols for the unknown and its powers, a major role is played by the concrete number symbols which double as parameters. In the latter case they can play the role of free parameters or of non-free parameters satisfying certain supplementary conditions.

Diophantus' algebraic formalism represents a *special stage* in the evolution of algebra which began with his "Arithmetic." This stage lasted in European algebra almost to the second half of the 16th century. It was only then that Bombelli and Stevin introduced symbols for a second, third, and further unknowns and Viète introduced symbols for parameters as well as the system of literal calculus.

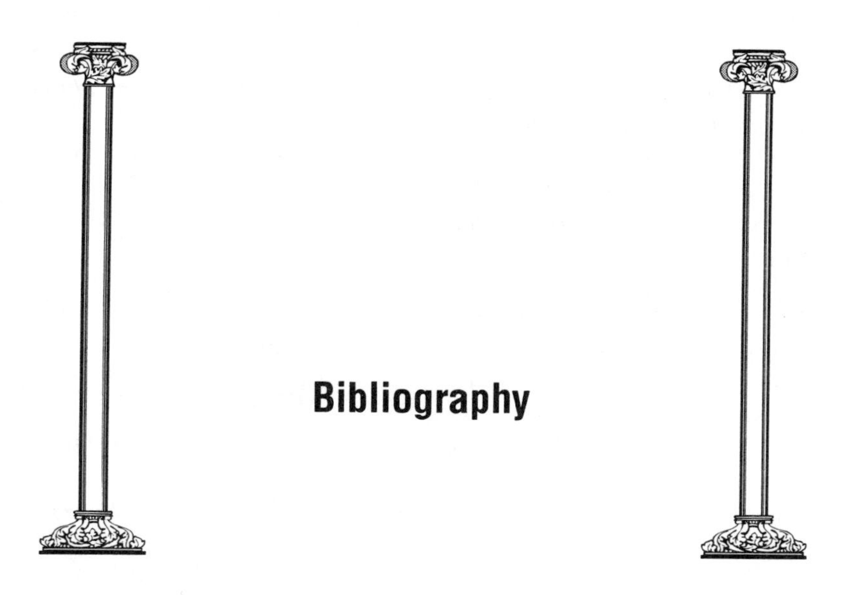

Bibliography

The following editions of the works of Diophantus are based on Paul Tannery's *Diophantus Alexandrini opera omnia.* Cum Graecis commentariis edidit Paulus Tannery, Lepsiae, 1893, vols. 1–2. They are:

[1] Th.L. Heath, *Diophantus of Alexandria,* A study in the history of Greek algebra. Reissued by Dover in 1964.

[2] *Arithmetik des Diophantos aus Alexandria.* Aus dem Griechischen uebertragen und erklaert von Arthur Czwalina, Goettingen, 1952.

[3] *Diophante d'Alexandrie,* Les six livres arithmétiques et le livre des nombres polygones, trad. par Paul Ver Eecke, Bruges, 1926 (reissued in Paris in 1959).

The following are historical works with substantial comments on Diophantus:

[4] D.J. Struik, *A Concise History of Mathematics* (New York: Dover, 1987).

[5] O. Ore, *Number Theory and Its History* (New York: McGraw Hill, 1948).

[6] B.L. van der Waerden, *Science Awakening,* transl. by A. Dresden (New York: Wiley, 1963).

[7] H.G. Zeuthen, *Geschichte der Mathematik im Altertum und Mittelalter* (Copenhagen, 1886; Neudruck, 1966).

[8] H. Wussing, *The Genesis of the Abstract Group Concept,* transl. A. Shenitzer (Boston: The MIT Press, 1984).

[9] J.E. Hofmann, *Geschichte der Mathematik* (Berlin: Walter de Gruyter 1953–1963; 3 vols.).

The following deal with diophantine equations and related number-theoretic issues:

[10] J.H. Silverman, *A Friendly Introduction to Number Theory* (New Jersey: Prentice-Hall, 1996).

[11] L.J. Mordell, *Diophantine Equations* (New York: Academic Press, 1969).

[12] J.W.L. Cassels and A. Froelich, eds., *Algebraic Number Theory* (New York: Academic Press, 1967).

[13] H. Davenport, *The Higher Arithmetic* (New York: Harper and Bros., 1963).

The following deal with group theory, with rational points on algebraic curves, with algebraic curves, and with algebraic geometry:

[14] J.H. Silverman and J. Tate, *Rational Points on Elliptic Curves* (New York: Springer, 1992).

[15] F. Kirwan, *Complex Algebraic Curves* (Cambridge, LMS Student Texts **23**, 1992).

[16] P.A. Griffiths and J. Harris, *Principles of Algebraic Geometry* (New York: Wiley, 1978).

[17] J.J. Rotman, *The Theory of Troups* (Boston: Allyn and Bacon, 1973).

The following deals with complex functions:

[18] G.A. Jones and D. Singerman, *Complex Functions* (Cambridge, 1987).

The following is a unique book, of special relevance for the Bashmakova book (translator):

[19] A. Weil, *Number Theory: An approach through history* (Basel: Birkhäuser, 1983).

Name Index